THE AGING GAP BETWEEN SPECIES

ANCA IOVIȚĂ

The Aging Gap Between Species

Library of Congress Control Number: 2015915887
CreateSpace Independent Publishing Platform,
North Charleston, SC

ISBN-13: 978-1517484811

ISBN-10: 1517484812

Disclaimer

Dedicated to each inquiring mind who added a small step to our understanding of aging

Contents

Finding the Forest Among the Trees

Aging is a puzzle to solve.

This process is traditionally studied in a couple of biological models like fruit flies, worms and mice. What all these species have in common is their fast aging. This is excellent for lab budgets. It is a great short-term strategy. Who has time to study species that live for decades?

But lifespan differences among species are orders of magnitude larger than any lifespan variation achieved in the lab. This is the reason for which I studied countless information resources in an attempt to gather highly specialized research into one easy-to-follow book. I wanted to see the forest among the trees. I wanted to expose the aging gap between species in an easy-to-follow and logical sequence. This book is my attempt at doing just that.

Aging is inevitable, or so I've been told. I was never one to accept things at face value just because some authority said it. So I began to question whether aging is the same in all species. While looking for answers, I was surprised to find out there is a lack of biological model diversity in gerontology. I was undeterred and I searched for the most obscure scientific papers on how other species age and what could set them apart. That's how I started typing the words you're now reading.

If you ever had a pet, you already noticed that lifespans differ widely. You may have looked the same for a decade, while your dog or cat already suffered from age-related diseases. There is huge lifespan variability, both in terms of individuals belonging to the same species and among species themselves. What are the mechanisms underlying the aging gap between species?

I intentionally chose to write the answer to this question in plain English. Aging research is too important to hide it behind the closed doors of formal scientific jargon. This book could not have existed if green tea, libraries and the Internet were not invented. The amount of data I had to browse in order to keep the essential patterns is huge. Yet this book is not exhaustive. This is not a dry academic textbook. I tried to instill life in a topic that is hugely important for the extension of human lifespan. Only you can decide if I achieved this.

Being Reliable Counts

Depending on the number of vital parts and the resilience of each of them, systems can range from the simplest to the most complex. Aging is first of all a consequence of being complex [42]. Here is why.

A system at its simplest has *one* vital part only. You could say bacteria are simple because they are organisms made of one cell only. If any external or internal factor kills that one cell, the whole system is disrupted and the organism is plain dead. As you'll read later, some bacteria display a form of aging, similar to clonal senescence in complex multicellular organisms, but I included them here because they are simple biological systems everyone is aware of.

When complexity increases from bacteria to a human being like you - made up of billions of cells with several vital parts -, aging becomes a reality. Time passes and each of these parts can suffer damage. Such damage is minor, but it all adds up. A blurry lens in the form of cataracts does not normally kill

2

people, unless they insist on driving cars or crossing the street unattended. Usually people with cataracts live for many years. But add heart failure to that. And minor kidney disease. No wonder a full-blown pneumonia infection can kill an elderly while leaving the young alive in most cases. Pneumonia may have been the final blow, but all minor and major damages a complex human being experiences lead to the final stage of involution: senility.

The Mathematics of Aging

Innovations often come from unexpected places. People searched for the fountain of youth since immemorial times, but it is only recently that aging itself received an *almost* mathematical description.

The solution came from people selling life insurance. If you want to make a living out of it, you need to correctly estimate the probability of individual X dying this year, next year and so on. If you miss the boat, you may undercharge him/her and soon get out of business. Overcharge people and they will soon prefer to set such funds on their own. In order to establish such a balance, actuaries started creating life tables and they noticed the probability of dying doubled each 8 years, with the lowest risk right around puberty. This is known as *mortality risk doubling time* in gerontology, the science of aging [34].

Humans are the only species to buy life insurance, but the pattern of doubling mortality risk stays the same in most other species, mammals or not. What differs is the time interval at which the probability to die doubles.

Apart from the mortality risk doubling, the other mathematical feature of aging is a drop in fertility [34]. This drop can take place gradually or almost abruptly like in female mammals undergoing menopause.

The aging process is [55]:

- progressive
- endogenous
- irreversible
- deleterious to the individual

Since it is in the interest of living beings to reproduce as many times as their bodies and the surrounding environment will allow, it is automatically in their interest to survive as much as possible. An organism does the best it can under the constraints of predators, insufficient resources and the desire to spread its genes in offspring that will survive too! According to the disposable soma theory, if an animal is predated upon during its midlife it makes no sense to divert unnecessary resources from reproduction towards better DNA repair genes that may never get the chance to be used [34]. But if the environment improves and animals reproduce at later stages in life, improving maintenance is a sure bet.

The Speed of Senescence

During the course of this book I use 'aging' and 'senescence' as interchangeable for the sake of simplicity. In gerontological literature though, *aging* refers to the passage of time without mentioning whether the system changed for better or for worse. *Senescence* describes the wear-and-tear common in most systems after days, weeks, months or decades depending on its reliability and expected lifespan. But when people casually talk about 'aging' , involution is usually implied.

The first major gap in comparative gerontology was closed when three patterns of senescence were recognized [34].

These put the rest of the book into perspective:

- rapid senescence
- gradual senescence
- slow or negligible senescence

Negligible senescence was the most difficult to accept by the gerontological community. All three are important for seeing the forest among the trees when it comes to aging itself.

Rapid senescence is spectacular. The *Oncorhynchus* Pacific salmon undergoes major damage to its organs right after spawning. Its aortic wall gets thicker. It develops fungus infections on its skin. The fish will most likely be dead in a couple of days. It is very rare for Pacific salmons to reproduce twice in a lifetime [34].

You may think rapid senescence would be an exception since evolution would have favored species that reproduce repeatedly. But rapid senescence does not impede multiple bouts of reproduction. The bamboo tree blossoms repeatedly for decades, but it stops doing so a couple of months before succumbing to death [34]. The bamboo's *fast growth* is a common theme in species with rapid senescence, hence its ubiquitous use in eco-friendly wood products.

Sometimes rapid senescence is manifested in one gender only. *Antechinus* marsupial male mice experience deadly exhaustion after mating with several females [34].

The common denominator of rapid senescence is mostly hormonal determination, especially the secretion of cortisol, a stress hormone. Other mechanisms are waiting to be discovered.

Gradual senescence is a milder form. Human beings are a good example. Aging occurs in decades which – given the average human lifespan – means that you spend about a third of your life growing, a third as a mature adult and the last third losing what you've built during the previous life stages. With few exceptions like the male marsupial mice mentioned earlier, most mammals undergo gradual senescence [34].

Slow or negligible senescence is where things become interesting. Such species display a constant mortality risk over their lifespan, while their fertility is constant or may even increase in time. In other words, they are *potentially* immortal.

An even more accurate classification of senescence according to mortality risk and fertility is the following [112]:

- positive senescence – where individuals of a species develop wear and tear signs as time passes by, their mortality increases and their fertility decreases

- negligible senescence – where aging signs are apparently lacking and the older individuals have the same probability of dying as the younger ones, while fertility remains the same

- negative senescence – where such individuals enjoy a diminished probability of dying as they grow older, while their fertility increases. As an example here, it is very easy to die as a one-day turtle, but once you reach the ocean and you are the size of an adult, the probability of dying is seriously diminished. Similarly, older male lobsters enjoy greater fertility than younger ones.

It took a long time until the second and third patterns of

6

senescence were accepted by the gerontology community. The status quo was that we were all destined to age no matter our genetic heritage. Whether one species belongs for certain to one of these three senescence phenotypes remains to be researched for most organisms.

Case Study: Aging in Fish

There is probably no other category of animals in which the speed of aging is so different. From the outside, fish look mostly the same. Their size may be different, but their anatomy is very similar. Fish are more homogeneous than birds or mammals. Yet their maximum lifespan ranges from a couple of months to more than 200 years and their rate of aging is just as different [87].

Fish which undergo **rapid aging** often breed only once. In other words, they are semelparous. Such species include:

- the *Oncorhynchus* Pacific salmon
- the *Anguilla anguilla* eel
- the *Mallotus villosus* capelin
- the *Petromyzontiformes* lamprey

Fish which undergo **gradual aging** include many species familiar to aquarium hobbyists:

- the *Poecilia reticulata* guppy
- the killifish group
- the *Danio rerio* zebrafish
- the *Oryzias* Japanese rice fish

- the *Xiphophorus* platy fish

Negligible aging fish grow slowly throughout their lifetime and have very low metabolic rates at old age. Examples include:

- the *Sebastes* rockfish
- the *Acipenser* sturgeons
- the *Huso huso* beluga
- the *Allocyttus verrucosus* warty oreo
- the *Polyodontidae* paddlefish
- the female plaice

Fish have plenty of predators. Yet relative to other vertebrates, fish often express delayed senescence because of indeterminate growth, a case where fecundity increases with age thereby favoring the older individuals in the total gene pool [93]. Whether these differences in aging speeds are due to hormones, telomerase or any other mechanism is still an open question.

How to Estimate Chronological Age

Birth certificates are a recent human invention. Most people nowadays are able to tell you their age and have a piece of paper to prove it too. Yet gerontology is about studying aging in other species as well. Except for certain pets, most of them don't receive any additional documentation at the moment of their birth.

This little piece of information is hugely important when determining the maximum lifespan of different species. Fortunately, gerontologists developed several age estimation methods.

Several species leave a trail of their *periodic growth increments* in the hard structures of their body. Trees display growth rings in their trunks, giving birth to the scientific method of dendrochronology or tree-ring dating. Bivalve shells archive their age in periodic growth lines. Corals display growth rings as well. Marine mammals have cementum layers in their teeth. Baleen whales have no teeth, but they grow ear wax plugs as they age [109]. Fish retain such growth patterns in their inner ear bones called otoliths, as well as in their vertebrae and scales.

The caveat is that growth patterns depend on local temperatures, seasonal food supplies as well as hormonal patterns such as reproduction. In order to be used as a method of age estimation, the above growth ring counting has to be *calibrated for each species*.

Such calibration can be done two ways:
- tag and later recapture individuals from the set species
- measure known radioactive decay in the hard structures and compare the result with the number of growth rings.

Only after calibration can you make sure whether a growth ring counts for one year.

Many plants and animals leave behind trails in the form of hard structures: shells, bones, teeth. But not all of them do that. So how do you determine the age of such an individual?

The tissue proportion of *lipofuscin*, a well known aging pigment, tells you the physiological age of that individual. Depending on the oxidative stress it encountered, this estimate may or may not coincide with the chronological age. For example, if you put under the microscope a sheet of tissue from someone suffering of lipofuscinosis, you may obtain a larger number of years compared to the one deducted from a healthy individual's identity card. Beyond humans, lipofuscin is deposited in several animal tissues.

Another estimation method could be the *telomere length* in somatic tissues. This method is reliable in species which *don't* express indeterminate growth by turning on the telomerase enzyme. As you'll see in the following pages, this method doesn't work in indeterminate growth species such as lobsters, crabs, sea urchins and some fish.

Taking Life Slowly

Entropy is commonly understood as a measure of disorder. A closed system with no external energy source has no choice but to reach maximum entropy. I often wondered whether aging is an example of entropy. Organisms need energy to at least survive, if not reproduce as well. Biological systems are open systems, hence according to the second law of thermodynamics, they can decrease their entropy if they increase entropy around them by at least that same amount. In other words, open systems can increase or maintain order if they create at least the same degree of disorder around them. I say 'at least' because energy is partly transformed in heat.

Entropy is directly proportional to energy and inversely proportional to temperature.

Depressing metabolism in different forms of **dormancy** like hibernation, estivation, diapause and many others is one way through which negligible and very slow senescence species buy themselves some time during which aging doesn't seem to take place.

The influence of temperature over longevity is one of the most interesting problems in gerontology. On an intuitive level, you learned to store your food in the fridge. The metabolic rate is decreased and food keeps longer without spoilage. It is intuitive to expect that low temperatures could decrease the rate of whatever metabolic pathways are involved in aging. Is this simple explanation enough for all species? Keep reading to find out!

On Temperature and Aging

The speed of most chemical reactions depends on temperature. Metabolic chemical reactions make no exception. Logic follows that decreasing the habitual temperature may prolong lifespan. Things aren't so simple though.

The first caveat is that temperature can't be dropped beyond a certain limit for an organism to survive. The limit is largely individual for each species and less so for each individual of a certain species.

The second caveat refers to the way organisms control their inner core temperature. When it comes to animals, heat management divides them in two main categories: cold-blooded animals or **ectotherms** and warm-blooded animals or **endotherms**.

Ectotherms rely on the external environment for providing them the right temperature range to survive. They will modify their *behavior* according to the available sources of heat. You may have noticed lizards basking in the sun.

Endotherms like us have internal heat management systems. For example, when you feel cold, your blood vessels constrict in order to prevent additional heat loss and your muscles generate heat by shivering.

The metabolic rate of ectotherms is a direct function of their surrounding temperature. There are two possible hypotheses here.

The first is that closely-related species living in colder environments will enjoy longer lifespans. Indeed, many fish and invertebrates live longer if located towards the poles and/or in deeper waters. For example, the ocean quahog *Arctica islandica* has a higher maximum lifespan around Icelandic waters compared to the German continental ones [105].

The second hypothesis is that bringing an ectotherm in a colder environment will prolong its lifespan and lower its oxidative damage, whereby a higher temperature will decrease lifespan.

- This was achieved in the lab for short-lived killifish which displayed a lower accumulation of lipofuscin as well [63].

- Raising the *Nothobranchius furzeri* fish in 22 versus 25 degrees Celsius increased its lifespan and decreased its lipofuscin accumulation rate [63].

- A decrease in temperature with a subsequent lifespan increase versus controls was noticed among individuals from the same species such as *Aequipecten opercularis* and *Adamussium colbecki* [63].

- On the other hand, the pearl clam *Margaritifera margaritifera* displayed accelerated senescence when living in warmer waters [63].

- A nonlinear increase in lipofuscin accumulation in higher temperature environments was noticed in crustaceans such as *Cherax quadricarinatus* and *Homarus gammarus* too [63].

The extreme longevity of certain ectotherms - like the *Arctica islandica* clam – may have evolved as a side effect of adapting to cold. When low temperatures set in for long periods of time, animals may respond by increasing their number of mitochondria or the number of inner membranes of these organelles [105].

Things aren't so simple in endotherms though. When encountering colder temperatures, such animals increase their metabolic rate in an attempt to keep their temperature in their normal range. If the exposure takes place for a long time, whatever that is for the species in cause, endotherms experience hypothermia from which they may recover more or less. Animals which regularly survive winter by hibernation are able to recover from the associated hypoxia and ischemia often induced by prolonged cold without suffering from reoxygenation-reperfusion injury.

One of the effects of calorie restriction with optimal nutrition is a drop in core body temperature. This change takes place in endotherms too. It's like there is a lower shift of 'normal' body temperature, often associated with a drop in metabolism reflected in lower thyroid hormone T3 levels.

A drop in consumed food during calorie restriction will often lead to a drop in core body temperature. Could we switch these two steps? Could we drop the body temperature with drugs, thereby prolonging lifespan? Unfortunately, the study that could answer this question hasn't been done yet.

Invertebrates are a group of species where lowering temperature on the long term has visible effects on prolonging lifespan. Invertebrates are ectotherms. In other words, their core body temperature largely depends on the surrounding temperature. A drop in ambient temperature leads not only to a longer lifespan, but to a slowdown of development and life history in general.

Here are a couple of species where a drop in environmental temperature was correlated to an increase in lifespan [63]:

- the *Drosophila melanogaster* fruit fly
- the *Caenorhabditis elegans* worm
- the *Trichogramma platneria* wasp
- Antarctic sponges such as *Cinachyra antarctica* and *Scolymastra joubini*

Moving up the complexity scale, **cold-blooded vertebrates** show the same pattern of life prolongation with lowering environmental temperature. Such vertebrates include fish and amphibians and examples include [63]:

- the *Sander vitreus* walleye
- the *Cottus bairdii* mottled sculpin
- the *Austrolebias adloffi* fish
- the *Nothobranchius furzeri, N. rachovii, N. guentheri* killifish
- the *Ambystoma macrodactylum* salamander
- the *Rana aurora* frog
- the *Proteus anguinus* olm

Unlike the rest of the species mentioned in this list, the olm displays signs of negligible senescence [103; 113].

When it comes to **warm-blooded vertebrates** like mammals and birds, the story becomes more complicated. Such animals maintain their temperature range between narrow limits.

Consequently, the neuroendocrine systems making this possible play a big role in accelerating aging.

The main gerontological intervention leading to a drop in core body temperature in **mammals** is calorie restriction with optimal nutrition. The drop in temperature is not spectacular though [63]:

- 1-1.5 degrees Celsius in rodents

- 0.5-1 degrees Celsius in rhesus monkeys

- 0.2 degrees Celsius in human volunteers practicing this regime between 6 months to 6 years

Calorie restriction with optimal nutrition leads to specific energy conservation adaptations like a decrease in circulating T3 hormone levels which further lower body temperature. Other hormone decreases with a secondary effect on body temperature are those of insulin, leptin and total testosterone [63].

So when it comes to complex animals like mammals and increased lifespan, it's difficult to separate the effects of lowering core body temperature from the effects of derived physiological adaptations. In other words, low temperature may be a stressor that leads to enhanced survival, just like calorie restriction *per se* is.

Actually, two confounding factors that may lead to the erroneous explanation that low temperature itself is a longevity factor are fewer predators in harsh, cold environments as well as fewer microorganisms responsible for infectious diseases.

I already mentioned that depending on the location of their energy source, animals can be endothermic or ectothermic. But when it comes to the degree of inner body temperature variation, they can be homeotherm or poikilotherm. The ability of some animals to maintain a *stable* body temperature despite environmental temperature variations is known as **homeothermy**. The opposite of homeothermy is **poikilothermy**, in other words the ability of some animals to vary their body temperature along with the environment's temperature.

Most endothermic animals are homeothermic. Animals with *facultative* endothermy are mostly poikilothermic. The sustained energetic output of an endotherm is higher than that of an ectotherm, but endotherms can only function over a narrow range of temperatures.

A special case of homeothermy is made up of those species entering *torpor* or *hibernation* during extreme temperatures. The inner core body temperature drops and the animal switches from endothermy to ectothermy. When it comes to comparing mammals of the same size, those hibernating reach a 50% increased maximum lifespan compared to the non-hibernating ones [110]. Such hibernating mammals are able to space reproduction events wider, leading to lower reproduction rates and longer generation times, a common feature in species with slow aging. Hibernation may have evolved as an adaptation to avoid predation. Not only does the animal stay put, hereby avoiding direct contact with predators, but the decreased metabolic rate leads to reduced detection by its enemies [110; 63].

In mammals, poikilothermy is mostly a feature of hibernating animals with one exception: the non-hibernating naked mole rat. It is difficult to mention an average body temperature for an animal at the whim of environmental temperatures, but when it comes to the naked mole rat its usual temperature ranges from 30.6 to 34.2 degrees Celsius [63]. There is an increase of 1.5 degrees Celsius during pregnancy, but this doesn't affect the longevity of queens [63]. Mole rats as a group have a lower body temperature compared to the rest of the rodents [63]. The naked mole rat is the mole rat with the lowest core body temperature and the longest living one too. Its low body temperature may be a factor in its longevity, but it's not the only one as you'll see in the following pages.

The longest living mammal, the *Balaena mysticetus* bowhead whale has the lowest metabolic rate among cetaceans, as well as a core body temperature of around 33.8 degrees Celsius, much lower than expected for non-hibernating non-marsupial mammals [63].

Here are a couple of **outliers** where having an increased temperature is associated with a higher lifespan [63]:

- women have a slightly higher body temperature, yet live more than men

- when comparing birds and mammals of same size, birds have a higher core body temperature and enjoy a longer lifespan

- the *Sciurus carolinensis* Eastern gray squirrel has an average core body temperature of 38.7 degrees Celsius, yet it survives around 23.6 years compared to the average rodent at 36.8 degrees Celsius with a 9.2 year expected lifespan

Dormancy

On the way to increase their survival and reproduction efficiency, species learned to temporarily pause the movie of life without stopping it altogether. More than this, they learned two strategies to synchronize the entry into dormancy with their environment. The organism may enter dormancy *before* adverse conditions using **predictive dormancy**. Otherwise, the organism may enter dormancy *after* adverse conditions started by using **consequential dormancy**. The latter gets into action when environmental conditions are highly variable, hence impossible to predict. The former may allow organisms to be active as much as possible before the worse begins.

The main advantage of dormancy is that organisms are able to schedule their reproduction in times of plenty when their offspring stand the biggest chance of reaching maturity. Dormancy is mainly triggered by changes in the daily amount of light, food availability and temperature variations.

Seed dormancy is the ability of seeds to prevent germination in unsuitable environments. This ability is crucial to the survival of the species. It's better to wait for suitable times when the probability of seedling survival is higher.

The same strategy is used by many other species. If you had children long after puberty set in, then you used it too. Delayed gratification pays off in people. Delayed germination pays off in plants.

Diapause is a type of predictive dormancy whereby the development stage is delayed to reach better times. It is common in worms and insects. Diapause is regulated by environmental triggers such as drought as well as certain neuroendocrine signals such as the juvenile and the prothoracicotropic hormones. A particular type of diapause in mammals is delayed implantation or embryonic diapause, where the embryo does not immediately implant itself in the uterus wall. Little or no development takes place during this stage.

Examples of consequential dormancy include **estivation, brumation** and **hibernation**. High temperatures and arid conditions trigger estivation in many species such as snails, insects, crustaceans, salamanders, tortoises and even the *Protopterus aethiopicus* African lungfish [28]. Brumation can be encountered in reptiles which can go for months without food at the set of fall. Hibernation is a case of dormancy in endotherms being able to switch to ectothermy. Hibernation refers to animals that employ switching between poikilothermy and homeothermy as well. Many of them are mammals like rodents, bats, bears and the fat-tailed dwarf lemur of Madagascar. Rodents employ several survival strategies: migration, food hoarding, hibernation and sometimes estivation too. The lifespan difference in rodents that hibernate and those that don't is more than 10 years for the hibernating ones [110]. During hibernation such animals decrease their temperature, heart beat, breathing rate and many of their hormone levels. All these examples allow species to save on the energy they may have employed to maintain their regular temperature.

A more down-to-earth type of depressed metabolism is **sleep**. Species differ in their sleep duration just as much as they differ in their maximum attainable lifespan.

Animals adapted their sleep to their external environment. Species having to escape many predators evolved unihemispheric sleep whereby only one cerebral hemisphere rests at a time while the other stays wide awake. Taking it a step further, species with almost no predators can afford to sleep a longer amount of time.

Consuming plants takes a lot of time, hence less time for sleep itself. Herbivores sleep less than carnivores, even when those herbivores don't have predators because of their sheer size. Elephants sleep around 3 hours. Lions sleep around 13 hours.

When compared to invertebrates and simpler vertebrates, endothermic mammals and birds have an added somatic cost in maintaining their inner core body temperature constant [100]. Given the same mass, cold-blooded animals have a lower metabolic rate compared to warm-blooded animals like us. Their food consumption is lower too. Since REM sleep brings a regular endothermic animal to poikilothermy, it is expected to encounter it as an adaptation in warm-blooded animals only. Compared to wakefulness, the metabolism drops during NREM too, but thermoregulatory processes can still take place. In other words, you can shiver or sweat during NREM sleep, but not during REM sleep. Thermoregulation is suspended during REM sleep, but heat production is not.

The metabolic rate is the same during NREM and REM stages. Even if thermoregulation doesn't take place during REM sleep, the energy saved on maintaining a constant inner core body temperature is spent elsewhere on other biological processes.

The amount of time an animal spends during REM sleep, as well as the number of cycles alternating from REM to NREM and vice versa is highly variable in different species. Generally, predators have longer periods of REM sleep compared to prey. As body size and brain size increase, so does the amount of REM time for that species.

Yet that is not the case for the total amount of sleep. The latter increases in smaller-sized mammals and birds. Presumably by employing REM sleep, animals manage to save the energy it would take to maintain their inner core body temperature and allocate it to better uses in the future.

Smaller animals have a higher surface-to-volume ratio and they easily lose heat, hence they expend a huge chunk of their energy in maintaining a constant inner core body temperature and need more sleep to recover [100]. Larger animals have longer REM bouts and longer REM-NREM alternating cycles during their sleep [100].

21

Although sleep is regarded as important in memory consolidation, some of the most intelligent marine mammals, cetaceans, don't experience REM sleep. The amount of REM sleep decreases with age in human beings. Moreover, species whose offspring are born very immature share a larger percentage of REM sleep compared to those where the offspring are more autonomous at birth [100].

The restorative function of sleep was proposed many decades ago. Indeed, in humans at least, hormones mainly secreted at night like growth hormone are anabolic ones, while those mainly secreted during the day like cortisol are catabolic ones [100].

Humans may become poikilothermic in acute diseases when the ability to regulate one's temperature is lost. Humans are homeothermic during wakefulness and NREM sleep. They switch to being poikilothermic during REM sleep [100].

It is not proved that all animals sleep. Some species may employ other strategies of allocating energy such as prolonged wakefulness, daily torpor or hibernation [100]. The energy savings of torpor and hibernation are even higher compared to sleep and that may be an adaptation when food supply is low. But when food exists and needs to be foraged, sleep is a worthwhile energy cost to pay in order to consolidate memories about where food can be found.

There is a huge group of organisms selected to survive hostile environments by periodically or completely halting their metabolism and getting into *suspended animation*. Once the environment gets safer, they wake up from their prolonged sleep and resume their normal lives. These masters of survival are

extremophiles. The most enduring ones are *Tardigrades* or water bears. These microscopic animals can survive hypersaline, high pressure, low pressure, high temperatures and freezing ones, severe droughts, shots of powerful radiations and as if all these wouldn't suffice, they are known to survive in void for 10 days [49; 60].

Ranging from cacti to bears, other animals and plants periodically decrease their metabolism as dry summers or freezing winters force them to. Since they slow down their metabolism, they are able to prolong their lifespan in absolute terms.

When the environment is scarce in resources, species inhabiting it have two solutions:

- either **reach maturity as fast as possible** and then produce **the maximum number of possible offspring**

- or **depress the metabolism** until the environment stressor will go away

The first strategy is used by short-lived species like mice.

The second one is commonly used by negligible senescence species such as the *Arctica islandica* clam, several species of turtles (*Chrysemys picta*, *Emydoidea blandingii* and *Terrapene carolina*) as well as the olm [103; 105; 22; 23; 77; 113]. The ability to undergo **dormancy** allows such species 'to sleep on their problems'.

In the struggle for life, many species combine decreased metabolism of their own bodies with storing supplies in remote locations for rainy days - squirrels hibernate AND bury nuts. If they wouldn't do both, their lifespan would be much, much shorter (although they still undergo aging wear and tear!).

Now suspended animation is a great mechanism of survival, but it won't make you young. It only defers the unavoidable.

The Housekeeping Problem

For many years a popular theory appeared in the field of gerontology stating that the full lifespan of a certain species is directly proportional to its metabolic speed or its heart rate. The classical example was the number of heart beats of the elephant compared to that of the mouse. Yet species with high metabolism such as bats and birds live more than expected, so this theory fell into disgrace. Aging is not simply burning up your reserves, but the theory contains some grain of truth in it.

Theories of aging belong to one of these two categories: **evolutionary** or **mechanistic**. Broadly, evolutionary theories of aging try to answer the 'why?' and mechanistic ones the 'how'.

The best known **evolutionary theory** is **the antagonistic pleiotropy or the trade-off theory**, stating that what gives an advantage when young will turn up a double-edged sword, as the organism tries to balance a limited amount of functional reserves between its own self-preservation and the growth of its offspring.

Mechanistic theories often revolve around the accumulation of oxidative stress damage in older individuals. Whereby a young individual starts out with a good balance between oxidative stress and cell repair mechanisms, it often undergoes more oxidative damage and/or fewer repair mechanisms as time goes by.

You know very well that if you don't regularly clean your house, clutter and garbage will accumulate. When you finally take your time to do housecleaning, you'll feel overwhelmed and maybe even procrastinate which will start a vicious cycle. This is exactly what happens in the aging body.

Older cells often accumulate intermediate and end products of oxidative damage. Such products can be grouped in:

- lipid peroxidation products such as ketones, aldehydes and fluorescent age pigments, including lipofuscin

- carbonylated proteins

- DNA damage markers

In short: lipid, protein and DNA damage increase with (physiological, not necessarily chronological) age. This accumulation is especially dramatic in postmitotic cells which are not able to *dilute* this material by continued cell division.

Oxidative damage takes its toll on all of us, but long-lived species often show less oxidative damage and/or increased resistance to oxidative stress compared to closely related short-lived ones.

Unlike humans who suffer from reoxygenation-reperfusion injury after prolonged times of hypoxia and ischemia, several species developed an extreme tolerance to anoxia: the *Arctica islandica* clam, the naked mole rat, the bowhead whale, the olm, turtles and many others.

Case Study: Aging in Turtles

Evolutionary speaking, turtles are one of the most conservative species. It seems like their survival strategy works. They avoid many predators by secreting a hard shell.

There are around 330 species of turtles, half of them being threatened with extinction. They often live more than 100 years. Turtles are reproductively active at very advanced ages. In most turtle species, the gender of their offspring is determined by the temperature at which they incubate.

Several species of turtles are considered negligible senescence species:

- the *Terrapene carolina* common box turtle [77]
- the *Emydoidea blandingii* Blanding's turtle [22]
- the *Chrysemys picta* painted turtle [23]

The Western painted turtle *Chrysemys picta bellii* is the most extreme anoxia-tolerant tetrapod known. It displays a very slow rate of senescence and like most such species, it is tolerant to extreme anoxia and partial freezing. If its water content is up to 50% frozen, the turtle can thaw and recover without any serious damage [104]. Humans would be long dead if placed in the same environment.

This turtle is able to survive on anaerobiosis for months. Presumably, the *Chrysemys picta* turtle is able to avoid the effects of lactic acidosis, the natural by-product of anaerobiosis, by its:

- blood composition

- shell

Namely, the calcium plasma levels increase during anoxia. Calcium is the buffer that reacts with lactic acid by forming calcium lactate, a precipitate which is stored in the bones and the shell of the animal keeping it out of the blood's way [56].

Anoxia tolerance is useful in minimizing oxidative damage. This can wreak havoc when leading to the accumulation of intracellular junk, the topic of the next section.

Intracellular Junk

As members of a gradually aging species, humans develop housekeeping issues as time passes by. Cells accumulate many types of intracellular junk that they can't recycle anymore. Just like it becomes more and more difficult to clean the house as decades pass by, it becomes even more difficult for cells to dispose of all their intracellular junk. They literally become cluttered.

This takes place at a faster rate in people who already introduce too much junk in their bodies, yet it happens in all of us nevertheless. Several substances accumulate, the best known being **lipofuscin**, also called **the pigment of aging**. Macroscopically we can see it in people with *age spots* – familiar, isn't it?

So why can't cells simply recycle this mess like they do during youth? The responsible organelle for intracellular recycling is the lysosome.

Its tools of the trade are [26]:

- specific enzymes

- a specific acid pH

As time goes by, lysosomes either become *dysfunctional* or they were *never adaptable enough* to deal with this kind of junk. Nature doesn't invest foolishly. By mid-life, most people were already dead because of predators, starvation, accidents, so *the human species didn't develop better lysosomes during its evolution*.

Lipofuscin accumulates in all of us if we are lucky enough to reach longevity, but in certain people this accumulation happens much faster. In *lysosomal storage diseases*, several types of intracellular junk accumulates at a faster rate and at a younger age because such patients lack certain enzymes or have dysfunctional variants of them. Meanwhile, healthy young people are able to deal with it and don't develop their symptoms and signs.

As regards senescence, there are three types of diseases where cells become choked with intracellular junk. They all increase in incidence and prevalence with age [26]:

- atherosclerosis
- neurodegenerative brain diseases
- age-related macular degeneration

Lipofuscin is deposited in the colon as melanosis coli, a condition often noticed during colonoscopy in patients with symptoms of constipation [37].

All these diseases may be slowed down by lifestyle, some of them can be 'managed' with drugs, but none of them can be cured. Until novel therapies are brought into practice – and even afterwards – it still pays to avoid putting more junk than necessary into your body by not smoking, by avoiding working in toxic environments and of course, by eating less in the first place!

Dr. Aubrey de Grey had the idea of searching for soil bacteria which may already contain **the necessary enzymes to break down lipofuscin** [26]. A couple of such species were identified and the specific enzymes must now be identified.

Caveat: lipofuscin is **not** the only substance humans accumulate with age, so several other enzymes may be necessary once the lipofuscin-breaking ones are identified!

Still, studying lipofuscin is an excellent start. Crustaceans are a group of species which seems to differentially accumulate fluorescent aging pigments.

Case Study: Aging in Crustaceans

Crustaceans lack permanent shells. They grow calcified structures, but they shed them with each moulting season. This makes it difficult to archive their chronological age, so researchers need to look elsewhere. They analyze the metabolic byproducts to estimate the age of a crustacean. Such products mostly include **fluorescent aging pigments**. The caveat is that such a strategy will rather determine physiological age compared to the chronological one.

When it comes to crustaceans, the amount of fluorescent age pigments increases with age, just like it does in humans. Their cells get flooded with intracellular junk such as lipofuscin. Yet the peculiar thing is that this accumulation of aging pigments with age takes place differently for each species. The accumulation rate in crustaceans can be linear, exponentially accelerating with age or decelerating with age [1].

Exponential accumulation with age takes place in species like:

- the *Callinectes sapidus* blue crab
- the *Homarus americanus* American lobster
- the *Aristeus antennatus* red shrimp

Linear accumulation with age takes place in species such as:

- the *Homarus gammarus*, *Panulirus cygnus* and *Panulirus argus* lobsters
- the *Pacifastacus leniusculus* crayfish
- the *Waldeckia obesa* amphipod
- the *Cancer pagurus* brown crab

Decelerated accumulation with age was noticed in:

- the *Cherax quadricarinatus* crayfish
- the *Penaeus japonicus* prawn

An exponential accumulation of lipofuscin with age may be due to a decrease in antioxidant defense as time passes by. In species with a linear accumulation of aging pigments, oxidative stress takes place at a constant rate during the animal's lifespan. A decelerated accumulation with age may be due either to increased cellular repair or to "dilution" in fast growing tissues.

As you'll see in a future chapter, some crustaceans undergo indeterminate growth. This may explain the differences in their accumulation rate of intracellular junk. Until then, extracellular junk is the topic of the next section.

Extracellular Junk

DNA codes *genes*. These are the blueprint of the myriad *proteins* it takes to build cells and make them interact with the environment and with each other. The sequence of the amino acids forming a protein determines its final 3D structure which largely sets its future activities. These macromolecules exist for a limited period of time. Later on they are degraded and largely recycled. I say *largely recycled* because mishaps do happen. Proteins become misfolded and the cellular machinery is damaged both in *accomplishing* its set activities as well as in *recycling* the damage.

There is a fine balance between *protein synthesis*, *protein folding* and *protein degradation*. Aging disrupts all this. Several proteins become misfolded with age, gathering themselves in sticky, insoluble aggregates, among which the best known is **amyloid**.

If you live long enough, amyloid aggregates will be found among your cells – right in the interstitial space. Such accumulations can take place at higher rates, sometimes starting during mid-life, in a group of diseases called **amyloidoses**.

Since many of them become more prevalent with age, it is *debatable* whether they represent an example of accelerated or normal gradual aging.

The most encountered forms of age-related amyloidoses are [26]:

- Alzheimer's disease
- cerebral amyloid angiopathy
- senile cardiac amyloidosis

Late-onset type 2 diabetes *may be* a similar disease since **amylin aggregates** accumulate in beta pancreatic cells responsible for insulin secretion [26].

The way species maintain the quality of their proteins plays a huge role in the extent of extracellular junk damage.

Case Study: Protein Quality Control

Compared to short-lived species, a feature of long-lived ones is their improved proteostasis [90]. Increased proteostasis leads to increased protein turnover and better cellular homeostasis.

Protein quality control is achieved by [90]:

- autophagy when removing and recycling long-lived proteins
- the ubiquitin/proteasome activity when removing short-lived proteins
- heat shock chaperones for achieving a stable state during the unfolding and refolding of proteins, so they don't oligomerize and aggregate

Here are six species of similar size with different expected lifespans belonging to three groups of animals like rodents, bats and marsupials [90]:

- the *Mus musculus* mouse 4 years
- the *Heterocephalus glaber* naked mole rat 30 years
- the *Nycticeius humeralis* evening bat 6 years
- the *Myotis lucifugus* little brown bat 34 years
- the *Monodelphis domestica* lab opossum 5 years
- the *Petaurus brevicaudus* sugar glider 18 years

Autophagy is enhanced in all the long-lived species mentioned above. Increased proteasome activity is present in long-lived rodents and marsupials, but not in bats. Heat shock chaperons are elevated in all three long-lived species, although the response to it is elevated only in long-lived rodents and marsupials.

It looks like the proteasome system and the heat shock chaperones didn't correlate with longevity in bats. Unlike the other two categories of species used, bats encounter huge temperature differences with maximum values during flying and minimum values during hibernation, so different stressors are probably needed to induce such protein quality strategies [90].

Autophagy (or self-eating) seems to be the common denominator in the aforementioned long-lived species.

There are three autophagy pathways [70]:

- macroautophagy
- microautophagy
- chaperone-mediated autophagy

Macroautophagy is absolutely necessary for removing whole damaged organelles, while microautophagy recycles material inside the lysosome.

As a side note, autophagy is negatively regulated by the mTOR pathway in mammals. Presumably, calorie restriction extends lifespan by activating autophagy [80]. It may extend lifespan by decreasing damage caused by glycation too.

The Sweet Poison

Just like a raw chicken gets crispy after spending two hours in the oven, especially if coated with a sweet sauce, so do human bodies "bake from within". This time the same process takes place in *decades* instead of hours.

Chemically this is called *the Maillard* reaction. Proteins mainly get cross-linked with sugars forming extracellular cross-links. These are called Advanced Glycation End-products or AGEs.

Humans are built of several proteins, most of which can become cross-linked during certain diseases and age stages. These extracellular cross-links lead to *stiffening* with age. Arteries become less able to react to blood pressure variations. Consequently, the elderly develop (mostly) systolic hypertension.

The crystalline lens quits *adapting* to variations in distance. Reading becomes difficult and by mid 50s almost everyone needs glasses for presbyopia. Later on, the lens becomes blurry with the onset of cataracts.

Normally, the *hemoglobin* protein is largely non-glycated.

Because of this, the percentage of glycated hemoglobin (*HbA1c*) out of total hemoglobin (Hb) is used as an indicator of the blood glucose levels during *the past 3 months prior to testing*. Since diabetes is characterized by a larger-than-normal average blood sugar level, it is reasonable to find higher than normal HbA1c percent levels in diabetic patients.

Diabetes is a classical model of accelerated aging. Here are its main complications. Among them, you will recognize many typical age-related diseases:

- cataracts
- macular degeneration
- diabetic retinopathy
- diabetic neuropathy
- kidney failure
- heart failure
- atherosclerosis
- silent heart attacks
- systolic high blood pressure

Diabetology as a medical specialty is one where prevention – whether primary, secondary or tertiary – is key to manage diabetes. Given the current state of medicine, once the patient shows clinical signs of AGE deposits, there is **nothing** that can be done anymore! Since diabetes produces no pain and largely no symptoms in the earlier phases, many patients don't understand the importance of medical treatment. They tend to view diabetes as a medicalization problem.

Lowering the amount of sugars and fats you eat is a healthy option, but you can't skip them altogether. Your brain needs glucose – preferably from vegetables and fruits. Your cell membranes need a decent amount of fat to survive. You need fat for maintaining the quality of your white matter in the brain and you need it to manufacture hormones. Does that mean cross-linking is inevitable? Is there any other way to at least slow it down?

It seems so. Researchers launched the hypothesis that *dietary* advanced glycation end products (dAGEs) may accelerate physiological protein cross-linking and they tested several food ingredients and several ways of cooking in order to gather a list of foods with their specific content of dAGEs [111].

Here are the main practical ideas to remember from the study:

- Animal products contain more dietary AGEs compared to plant ones, especially meat (no surprise here, meat is full of proteins).

- Marinating meat in acidic solutions – such as vinegar or lemon juice – before cooking it decreases the level of dietary AGEs once cooked.

- High temperature and low moisture types of cooking accelerate the formation of dietary AGEs. Avoid grilling and frying.

Among all the diseases that afflicted humankind, there is none that will age you as fast as diabetes. The simplest thing you can do today to slow down your aging is to *skip the sugar* and all its associated replacements.

Don't fool yourself that dried fruits are way healthier. Respect your body. You won't get a second one anyway. To stay on the safe side, introduce clean foods with *no* necessary list of ingredients to read. The simple things in life like vegetables, fish, eggs and fruits once in a while are always the best choice. Quitting the sugar addiction is difficult but just like quitting smoking, it is totally worthwhile!

Are Cell Membranes the Pacemakers of Metabolism?

One of the earliest observations in gerontology is the correlation of increased size with a longer lifespan. Presumably, being large makes one less prone to predators and allows better coping with starvation thereby decreasing your extrinsic mortality rate. At the same time, metabolic rate decreases when size increases and this could lower the extent of inevitable oxidative damage. This is known as the rate of living theory.

When comparing species among themselves, size *mostly* correlates with longevity. Yet a couple of exceptions like bats, birds and naked mole rats are able to survive much longer than expected given their size and increased metabolic rate [55].

Such exceptions have astonished researchers for years until a connection was made between the composition of cell membranes and the longevity of species. The animal cell membrane is actually a double layer of fats or lipids. Fatty acids count among the components of fats. They belong to one of the following three categories:

- saturated fatty acids
- polyunsaturated fatty acids
- monounsaturated fatty acids

Fat which is solid at room temperature is mostly saturated. The degree of fat unsaturation provides fluidity at room temperature because unsaturated fatty acids have a lower melting point compared to saturated ones.

The structure of fatty acids is largely dependent on temperature.

According to their structure, fats fall on a certain value of the peroxidation index which measures their susceptibility of being oxidatively damaged. Saturated and monounsaturated fatty acids are mostly resistant to peroxidation, while polyunsaturated ones get easily damaged [55]. Several end products of lipids peroxidation can further damage the cell membrane. Consequently, the proteins inhabiting the cell membrane will not undergo their normal functions and the permeability of the cell will vary. Being too lax about which substances enter the cell is just as damaging as being too restrictive and not allowing certain substances such as hormones to accomplish their tasks.

The permeability of cells and their organelles is determined by modulating the balance between saturated and unsaturated fatty acids. In biology, this is known as homeoviscous adaptation [101]. The strategy is especially important in poikilothermic organisms which can't regulate their own inner core temperature. Hence when temperatures drop, cell permeability is increased by maintaining a larger proportion of unsaturated fatty acids. When temperatures increase, saturated fatty acids dominate the picture.

The composition of cell membranes varies wildly among species and this may underlie their different metabolic rates [55]. The cell membranes of long-lived species are more resistant to peroxidation compared to shorter-lived ones. When it comes to mammals, body mass increases correlate with a decrease in polyunsaturated fatty acids in cell membranes. These fatty acids can't be synthesized *de novo*. They must be sourced from the diet or manufactured by local intestinal microorganisms.

The cell membranes of birds are more resistant to peroxidation compared to similarly-sized mammals [55]. Such differences exist between queens and their workers in eusocial insects despite starting out with the same genome [50]. All these may explain the longevity of organisms which don't fit the rate of living theory.

The next chapter will tackle the confusing connection between the onset of sexual maturity and the debut of senescence, especially when it comes to accelerated and gradually aging species.

Could Reproduction Set up the Pacemaker of Senescence?

Macroscopically, functional reserves deplete with age. On a smaller scale, mitochondria are not that efficient anymore.

When it comes to producing energy, determining genders and committing apoptosis or the programmed cell suicide, mitochondria secretly rule our lives [68]. Most human cells – which do have a nucleus, that is – contain two types of instructions: the nuclear DNA and the mitochondrial DNA. These types of DNA are quite different. They couldn't survive one without the other. Each cell has one nucleus and several mitochondria depending on its energetic needs, just like a car's engine has more or fewer cylinders depending on the type of car and the necessary speed it must develop.

Nuclear DNA is big and clumsy. In humans it is organized in 23 chromosome pairs. Some of the nuclear genes are absolutely necessary for mitochondria to burn energy by manufacturing ATP, our currency of energy.

Just like bacterial DNA, mitochondrial DNA is circular. Its genes are a lot more susceptible to mutations since many free radicals are produced within the mitochondria itself. When too many mutations are gathered and energy is lacking, the cell commits suicide in a noble fashion – and where does it all start? In mitochondria, of course.

Programmed cell suicide or apoptosis happens all the

time during human lifespan. It happens during development as some cells are only necessary during certain stages and are later committing suicide as their utility drops off.

It happens in infections too, especially when the culprit hides inside the cells determining them to manufacture foreign proteins. These microorganisms have learned how to bypass the complex immune system by hiding inside the cells. Yet the immune system evolved too and instead of relying on manufacturing antibodies, it now detects such cells as not really belonging to 'self'. When the cellular immune system detects such a malfunctioning cell, it makes it commit suicide [21]. Sometimes a whole organ is attacked, not so much because of the toxins synthesized by the microorganism, but because of the exaggerated immune system response. An example is the liver in hepatitis B.

There are two different types of DNA in most human cells. This is *a curse and a blessing* at the same time. Unlike prokaryote organisms, eukaryote cells produce high outputs of constant energy making all complex life possible. At the same time, there is a high risk of mutations. And with each new mutation, the frail harmony between the two DNAs is in danger, bringing chains of apoptosis as time goes by and making aging inevitable.

Have you ever asked yourself why does the human species have *two different genders*? Why not more genders? Why not just one? Why is sexual reproduction necessary? As far as we currently know, the mitochondrial DNA is usually inherited from the mother only. The sperm cell has mitochondrial DNA as well, but in normal cases, it is only used for fueling the cell on the way to the ovum. I say in normal cases, because there is a range of mitochondrial diseases where the same individual has inherited both mitochondrial DNAs.

Each somatic cell ages, yet the whole species doesn't age – isn't that odd? No matter how aged the parents may be, *the child will always be born young* – whether healthy or not. Somehow this ticking clock is reverted with each new human being. Each cell has a definite lifespan. In biology, this is known as '*the Hayflick limit*'. In other words, any cell cultured *in vitro* divides itself a certain number of times after which it stops. The cell does not immediately become dysfunctional, but its remaining days are numbered. This happens in germ cells too, hence fertility drops with age.

Each female fetus has millions of ova cells inside her ovaries. Once she is born, most of those germ cells would die leaving her with about 400 which will be selected once again at puberty. From all those initial million ova cells, she would now release about one ovum cell per month. Local natural selection among ova cells took place. The two types of DNA for each ovum cell are harmonious, allowing the cell to deliver energy and to sustain and maintain the growth of another human being if the ovum is fertilized [68].

In terms of a biological clock, the unfertilized ovum cell is not that different from a somatic cell. It will accumulate mutations and die just like the rest of the cells. What can such a cell do? It does what you do when you miss some information and find a new source of it – like a friend, a search engine or a book. *The ovum receives new DNA from the male sperm cell.* It receives nuclear DNA only [68].

Many combinations take place and some of these embryos will survive beyond the first pregnancy trimester. Spontaneous abortions are the rule rather than the exception. Given modern medicine, many of these fetuses will survive to be productive and happy grown-ups. The moral of this story is that it takes *many trials and errors* to turn back the clock and produce a new, young person out of aging cells [68].

Is aging a mortal sexually-transmitted disease? In other words, does all this mean aging is an inherent feature of sexual reproduction? Keep reading to find out!

The Segregation of Somatic and Germ Cells

It often seems like reproduction is the pacemaker of senescence in accelerated aging species. Postponing or avoiding reproduction in such species often increases the lifespan of individuals.

In gradually aging species, senescence takes more time to insinuate itself. But not having kids is not the fountain of youth. Monks and nuns age just as well. It is fascinating to study the link between reproduction and aging in gradually aging species like humans.

You have two different types of cells in your body: somatic cells and germ cells. Your germ cells are responsible for creating a new human being if you choose to reproduce. Your somatic cells are responsible for maintaining you before, during *and* after reproduction. In human beings these types of cells are segregated. This separation takes place early on during development. Not all species do the same.

There are two major factors influencing the phenotype of senescence in biological organisms:

- the timing when the segregation between germ and somatic cells takes place – during development or during the adult stage

- the type of reproduction – which can be sexual, parthenogenetic or vegetative, the latter strategies being asexual

With all these variations, species may exhibit the rapid, gradual or negligible phenotypes of senescence. No variation is exclusive to any type of aging.

As regards the segregation between germ and somatic cells, you may think this seriously impedes the ability of an individual to regenerate its tissues, but negligible senescence takes place both in [34]:

- cod and rockfish which segregate their germ and somatic cells during development
- as well as in conifers and hardwoods which segregate their germ and somatic cells as adults

The segregation of germ and somatic cells takes place in all three types of reproduction:

- sexual
- parthenogenetic
- vegetative

Sexually reproducing species like mammals and most insects do the segregation early on during development, while vascular plants and tunicates separate the two types of cells as adults [34].

Parthenogenetically reproducing species like a couple of lizards, insects, nematodes and rotifers segregate their cells during development, while the few angiosperms and ferns which replicate this way segregate their cells as adults [34].

A curious thing is that all vegetatively reproducing species – including corals, some flatworms and vascular plants – maintain separate germ and somatic cell lines once they reach the adult stage [34].

The preliminary conclusions I draw until now are:

- sexual reproduction and early germ-somatic segregation during development are compatible with negligible senescence as proven by rockfish and cod species.

- most species exhibiting negligible senescence segregate their germ and somatic cells during the adult stage as they are mostly species with vegetative reproduction. *Vegetative reproduction is impossible without maintaining the totipotency of somatic cells.*

- at the same time, many species like monocarpic plants and ascidians that differentiate their cells later on as adults exhibit rapid aging because of clonal senescence. *Vegetative reproduction is not the fountain of youth.*

Clonal Senescence Versus Mechanical Senescence

Most syndromes of aging are mainly determined by two types of senescence:

- mechanical senescence

- clonal senescence

Both types of senescence express an apparent limit in regenerating tissues after a certain biological, not necessarily chronological, age.

Clonal senescence is not only a feature of adult postmitotic cells. It can be encountered in vegetatively reproducing species too, such as monocarpic plants and ascidians [34]. On the other hand, these examples do not use vegetative reproduction exclusively.

Asexual animals maintain their telomere length constant, while sexual animals restore their telomeres only during embryogenesis. This extension of telomeres may be done with the help of the telomerase enzyme or, alternatively, independent of it.

Is clonal senescence related to the presence or absence of telomerase? I'll answer this question in more detail in the chapter on growth and aging.

Same Species, Different Lifespans

Apart from the type of reproduction and the onset of the somatic and germ cells segregation, the social role of an individual has considerable influence over its maximum attainable lifespan. Its social role may allow one to lead a safer life or on the contrary, be more exposed to the environment.

Such differences are visible in:

- eusocial species
- parasite/free-living populations of the same species
- individuals living in isolated environments like islands

These are the topic of the following three case studies.

Case Study: Eusocial Species

Ever since the discovery of the DNA molecule, researchers have been keen on separating the **genetic** influences on development and diseases from the **environmental** ones.

Human twins are rare. Studying them is no easy feat. At the same time, a huge number of twins are available for study right under the researchers' noses: the common **honey bees**.

All working bees in a honeycomb are *half twin sisters* sharing one mother and hence half of their DNA is common to the queen's. Through hormonal inhibition, the queen makes sure she is the only one laying eggs in the colony. Some of these eggs will make up the new generation of working bees, while a new queen will be chosen from special larvae [86]. The difference between all these female bees rests largely in their diet: all bee larvae are fed royal jelly during their first three days of life, but only the larvae destined to be queens enjoy royal jelly for the rest of their lives!

Long live the queen! Despite starting out with identical DNA, working bees survive a couple of weeks in the summer and up to several months in the winter, while the queen lives for about 3-8 years [86]. A new science was born studying the influence of the environment on the genes themselves: *epigenetics*. Many such **eusocial species** display this kind of differential lifespan as a consequence of the division of labor into reproductive and caste individuals.

Coming back to bees, a beehive can have no more than one queen. As soon as the first queen larva hatches from her hexagonal cell, she starts injecting poison in the neighboring cells killing off her competition. If more of them hatch out of their cells at the same time, a fight starts and the winner becomes the official queen, while the loser is either killed or leaves to set up another colony. Once the queen wins the fight, she comes back to her colony and if she is accepted by the female workers, she starts the nuptial flight, laying eggs for the rest of her life and being cared for by the female workers. She also starts secreting a hormone that impedes the ovulation of the female workers.

The preliminary conclusions are:
- the food these related female honey bees eat is very different and different nutrients – or a lack of them – may change the way DNA is expressed
- the queen bee acts more like a germ cell line in your body, while the female workers are inhibited from replicating by using hormonal inhibition just like somatic cells are repressed in adult mammals

Other eusocial insects where the queen has a longer lifespan than her workers are ants and termites [57]. Wasp queens are apparently an exception [86]. Eusociality exists in a couple of crustaceans from the *Synalpheus* group [31] and to a lesser degree in mammals.

A native of Eastern Africa, **the naked mole rat** is one of the few eusocial mammals. The members of the colony have their own social roles, yet unlike bees or ants, they can go up and down the ranks. In other words, they enjoy social mobility.

These little burrowing rodents feed on underground tubers only, hence they need to burrow all their lives. Not all individuals in a colony do this. There are specialized diggers, just like there are dedicated soldiers and a queen. When not needed, the soldiers rest in a central chamber underground to preserve their energy.

One or two of the eldest soldiers will mate with the queen when the breeding time comes. These dispensable soldiers risk their lives for the only indispensable individual of the colony: the queen. As long as she is alive, the queen is the only female who mates. She spreads her urine all over the colony and workers readily cover with it in order to be accepted by the rest of the members. Yet by being accepted, the rest of the females also accept sterility in exchange. The queen's pheromones preclude them from reaching sexual maturity.

Naked mole rats created a safe environment in which generation after generation they evolved their longevity and cancer-proof phenotype. Most rodents are easily preyed upon because they gather their food on the surface of the land. Some of them live underground, but their shallow burrow system is easily accessible to snakes. Naked mole rats build deep burrow channels and they seal them, creating their unique ecosystem [7]. This safe environment compounded with their social system gives them an edge in survival.

A deep burrow system is safe, but it also limits the nutrition of naked mole rats. They feed on tubers which regrow annually. Hence they have adapted to grow slowly, a feature of many negligible senescence species. Unlike other rodents, naked mole rats have one litter per year. They had to adapt to maximize their interval of sexual reproduction. Given their environment and limited diet, they couldn't afford to have many offspring per year like most other rodents. In their world, it pays to live longer. Given their limited annual growth, limiting reproduction to a couple of individuals paid off nicely.

Like all long-lived species, they are highly resistant to

stress. No such individual has ever been found to have tumors and this fascinating property will be reviewed in more detail in the subsequent chapter on cancer.

Case Study: Parasite/Free-Living Populations

Some of the most amazing things happen just before our eyes if we take a moment to notice them. Rats are disgusting creatures we avoid. Yet inside their intestines a parasite worm called **Strongyloides ratti** can be found. This worm is able to live both as a parasite and as a free-living creature during its life cycle. It can reproduce itself:

- asexually by parthenogenesis while living as a rat parasite
- sexually as a free-living creature outside the rat

The amazing thing about this shy worm is that two sets of female offspring can be created. Female parasitic worms produce female eggs only. Some of these go on reproducing parthenogenetically inside the rat's intestine while others leave the host infecting other rats and reproducing themselves sexually.

Although these two sets of female worms start out with the same set of genes, the maximum lifespan in the first case is **403 days**, while free-living adults live **5 days** at most [40].

These individuals are morphologically different, yet genetically the same. Why do individuals belonging to the same species display such an extreme aging plasticity? Which one is the 'real' lifespan of this worm?

Case Study: Island Versus Inland Populations

Genetically isolated populations are a lab designed by nature.

The *Didelphis virginiana* **Virginia opossum** is a short-lived marsupial spending its days in North America. It rarely survives more than two years. The animal may reach four years if living in captivity. Its mortality rate is so high that it never paid off to invest in biochemical mechanisms to prolong its lifespan. It has plenty of predators and very few defenses against them. Sure, feigning death and playing possum may help, but how many times does that work?

Yet a unique population of opossums developed a slower senescence phenotype on the Sapelo Island in the state of Georgia. Having no natural predators for thousands of years, the population of Virginia opossums slowed their aging, increased their average lifespan with 25-50% and had fewer offspring [4].

It seems like when there is an incentive to survive, nature will do whatever it takes and the average individual will increase its lifespan.

As you have read during the previous pages, there is no single type of reproduction that guarantees the existence or the avoidance of senescence. Vegetative reproduction is not a condition *sine qua non* of negligible senescence. Sexual reproduction does not always lead to the decline of physiological functions.

At the same time, the social role an individual plays has a moderate influence on its lifespan. This is true especially in those species undergoing mechanical senescence. With limited regenerative abilities, an individual may slow down the process of aging simply by avoiding certain tasks, like in the case of queen bees that have no need to fly except for their nuptial flight.

Hormones are the hidden conductors of the aging process, especially in the case of accelerated senescence species, so they are the focus of the next chapter – see you there!

Could Reproduction Set up the Pacemaker of Senescence?

Hormones as Pacemakers of Senescence

Hormones are a huge part of what aging looks like. You can't live without them, you can't live with too much of them and to make it more complicated, their normal thresholds change with age. It's like the rules of the game change while the game is still being played.

Nobody can say for sure where aging starts. Is it at the molecular level? Do cells age first and drag down the whole body? Is it a macroscopic form of disorganization that has top-down effects on all your cells and further to the molecules they contain?

The answer may be mixed. Yet hormones with or without the hypothalamus have an important role to play as **pacemakers of senescence**. The role of hormones is especially visible in species with accelerated aging.

Hormones are produced in several glands of your body. Each gland has three bosses. The first one is *the nervous system*. The second one is the community of *chemical receptors* found in the blood. As if two bosses were not enough, each gland is further controlled by hormones secreted by *other glands*. The purpose of this intricate hierarchy is to maintain the body's homeostasis. In other words, biological substances are maintained between narrow limits. That sets life apart.

Homeostasis is maintained as a child, as an adult and as an elderly as long as you are healthy. Yet each of those *developmental stages* has different *thresholds* beyond which homeostasis can't be maintained anymore.

I will further uncover the hormonal changes in human aging in order to better understand the subsequent case study in long-lived rodents.

As previously mentioned, there is a complex hierarchy of glands that produce your hormones and control your body. The part of the brain that indirectly controls major glands is **the hypothalamus**. As its name suggests, it is located right under the thalamus and close to the brain stem. It has plenty of nervous and endocrine tasks to accomplish.

The hypothalamus contains neurosecretory cells. These are neurons which secrete hormones. Such hard-working cells produce the following substances:
- the thyrotropin releasing hormone (TRH)
- the growth hormone releasing hormone (GHRH)
- the growth hormone inhibiting hormone (GHIH)
- the gonadotropin releasing hormone (GnRH)
- the corticotropin releasing hormone (CRH)
- oxytocin
- the antidiuretic hormone (ADH)

All these substances control **the pituitary gland** in a top-down fashion from the brain to the distal tissues. In humans the pituitary gland reaches its *maximum* size during middle age. Afterwards it gets smaller. There are backward feedback loops as well, where faraway cells communicate the local state to the brain through the hormones they secrete locally. It's all well orchestrated.

Even so, the quantity of hormones secreted by the hypothalamus remains constant. What does change is the *response* to each such hormone.

Through these hormones, the hypothalamus regulates your life in unforeseen ways. Here are the ways in which your hypothalamus influences the local activity in **the anterior**

pituitary gland:
- TRH secreted by the hypothalamus stimulates the anterior pituitary gland to release thyroid-stimulating hormone (TSH) which further controls the speed of your metabolism.
- GHRH and GHIH set the level of your growth hormone which has major implications in aging as you'll see further.
- GnRH influences your fertility by stimulating two tropic hormones: the follicle stimulating hormone (FSH) and the luteinizing hormone (LH), with FSH stimulating the follicle cells of the gonads to produce ova in females and sperm in males, while LH is responsible for the sex hormones secreted by the gonads themselves, estrogens in females and testosterone in males.
- CRH is the top hormone controlling your stress levels by stimulating the release of adrenocorticotropic hormone (ACTH).
- PRL or prolactin distally stimulates the breasts to produce milk during breastfeeding.

As people age, their water levels decrease and their proportion of fat increases. At the same time, loss of lean body mass is common, so there is less metabolically active tissue for the thyroid hormones to act upon. Since the whole metabolism decreases, it is common to see slight **TSH** increases in elderly without displaying any signs of hypothyroidism. TSH may increase with age because the incidence of thyroid autoimmune disorders like Hashimoto increases too [78].

We casually say that something has aged when growth stopped. In 1932 Bidder proposed *the hypothesis that aging starts when growth stops* [5]. Things aren't as simple as they seem. A cessation of growth is not necessarily associated with senescence and species with continuous growth may display aging signs like in the Guppy fish [92].

In aging humans at least, there is noticeably less growth as time goes by. **GHRH** and **GHIH** are both responsible for adequate secretion of the growth hormone (GH) which stimulates the tissues' growth, repair and cell division.

As time goes by, growth hormone is inhibited from being secreted. GHIH or somatostatin increases and the serum GH level decreases. At the same time, GHRH decreases with age. This has several consequences such as fewer daily peaks of GH secretion and probably sleep fragmentation which is so common in the elderly [78].

The hypothalamus communicates with **the posterior pituitary gland** too. Two of its hormones are transported, stored and released according to need: oxytocin and the antidiuretic hormone (ADH).

Since fertility drops after menopause, **oxytocin** with its roles in triggering uterine contractions and breastfeeding finds no use in the body anymore.

As its name suggests, the antidiuretic hormone (**ADH**) prevents diuresis or water loss. Because of this substance, your kidneys reabsorb water that could have been removed from your body and you lose less water through the sweat glands.

ADH is secreted inappropriately as years pass by, despite decreased blood vessel tonicity. Just like roses in a vase become dried after their blooming days, so do the elderly become easily dehydrated. To make matters worse, their thirst instinct is delayed and less intense compared to youngsters.
The aged kidneys develop partial resistance to the renal action of ADH, so many elderly develop an acquired partial nephrogenic diabetes insipidus [78]. In simpler words, they lose water even when their body water level is low to begin with. It's a vicious circle. Aged people drink less water and get to keep less of it in their bodies.

Posterior to the thalamus portion of the brain lies **the pineal**

gland. It is small in size, but big in effects. The pineal gland secretes **melatonin**, the hormone you need to know when to sleep and when to wake up. Melatonin is produced only in dim light or total darkness, because its secretion is inhibited by light reaching the retina of your eyes. The pineal gland is active at night, when increased melatonin levels make you feel sleepy. In nocturnal species the cycle is reversed with melatonin being secreted mainly during the day [100].

During human development, melatonin secretion peaks during early childhood and then it keeps decreasing to negligible levels in the elderly. An altered circadian clock is one of the causes of poor sleep in the elderly [78].

The butterfly-shaped **thyroid gland** is wrapped around the anterior part of your neck. It produces three major hormones:
- calcitonin
- triiodothyronine (T3)
- thyroxine (T4)

The thyroid gland undergoes moderate atrophy with age. It develops multiple nodules. Fibrosis is common. Many of the hormones it produces are modified too. Although the thyroid gland produces about 30% less T4 in the elderly, its serum level remains constant. This could be a compensation for the decreased use of T4 by the tissues, since elderly people have decreased lean body mass. Like many other substances in the body, the T4 thyroid hormone is cleared less efficiently and in consequence the pituitary gland refrains from stimulating the thyroid gland as it used to. The net sum of these effects is a constant serum value of T4 no matter an individual's age. With age, the thyroid gland produces less T3 and the serum level decreases as well [78].

Calcitonin is released when the body detects its calcium ion levels in the blood are increased. Consequently, the hormone leads to the absorption of calcium by the bones.

On the posterior side of each thyroid gland lobe you have **the four parathyroid glands**. They are responsible for producing the parathyroid hormone (**PTH**) which maintains the homeostasis of calcium and phosphate.

The parathyroid hormone gets into action when the body senses a drop in blood calcium levels. The hormone releases more calcium from the bone stores. It also influences the kidneys to reabsorb calcium instead of eliminating it.

The parathyroid glands produce more PTH with age. Consequently, calcium is released from the bones at an accelerated rate [78].

Osteoporosis was considered a normal part of aging decades ago. It is now considered a disease, maybe because there are treatments that decrease the risk of fractures if falls happen. There is an inherent link between the rate of osteoporosis and the rate of atherosclerosis. It's like calcium switches places from where it should be to where it shouldn't be.

Calcium is metabolized differently with age. Although absolute serum calcium levels stay the same in young and old people, there is a relative decrease in ionic calcium. This takes place because of less calcium intake, less calcium absorption and increased calcium loss.

Calcium is less absorbed because of the wide-spread vitamin D deficiency with age. Elderly people have less sun exposure, less vitamin D intake and absorption as well as less activation of D3 in the kidney and increased tissue resistance to the effects of this vitamin.

Most calcium in humans resides in the bones. It is only under controlled conditions that it is released into the bloodstream where it acts as a signal molecule. Once in the blood vessels, calcium either stands alone or it is bound to proteins. Just like macroscopically calcium is deposited in the bones, microscopically it is deposited in the mitochondria and the endoplasmic reticulum.

The metabolism of calcium is impaired not only in osteoporosis, but in atherosclerosis too as ectopic calcium is deposited inside arterial blood vessels and heart valves. Brain sand or corpora arenacea is commonly noticed on x-rays as calcium salts are deposited in the pineal gland and the choroid plexus.

The thymus is a soft gland located right behind your sternum bone. The hormones it produces, **thymosins**, train young T lymphocytes to make the difference between self and non-self. This gland becomes inactive during puberty. As people age, its tissues are gradually replaced by fat. Thymus atrophy may be one of the causes of immune senescence in the elderly [78].

On top of your kidneys lie your **adrenal glands**, each made of two different layers: the adrenal cortex on the outside and the adrenal medulla on the inside.
The adrenal cortex regulates your sugar, minerals and sex hormones through the hormones it produces: *glucocorticoids*, *mineralocorticoids* and *androgens*.

Glucocorticoids increase the level of sugar in your blood by breaking down other substances from which glucose can be released, such as fats and proteins. They reduce inflammation and because of this side effect, they are largely used in managing allergies, autoimmune disorders and transplant rejection.

The glucocorticoid name comes from a combination of three words: glucose+cortex+steroid. In other words, this hormone controls glucose, it is synthesized in the adrenal cortex and it has the chemical structure of a steroid.

Mineralocorticoids on the other hand are responsible for the homeostasis of the countless mineral ions inhabiting your body.

Androgens are produced in low quantities by the adrenal cortex to control the cells which are receptive to male hormones.

The internal layer of the adrenal glands, the adrenal medulla, produces two vital hormones with strange names: *epinephrine* and *norepinephrine*, also known as *adrenaline* and *noradrenaline*. These hormones can save your life by activating the 'fight or flight' response to stress. The hormones increase your heart rate, breathing rate and blood pressure. They increase blood flow to the organs that matter in acute stress such as the brain and muscles and decrease it in the organs that can wait such as the digestive system.

I don't know whether the aging stage of life is more stressful than others, but elderly people respond differently to stress as shown by their stress hormone level changes. Older tissues are highly resistant to the action of norepinephrine, hence its increased secretion with age. Epinephrine levels are normal or slightly decreased. Cortisol levels increase with age.

Hidden inside your abdominal cavity lies a gland whose influence on your sugar won't be noticed unless it's (almost) too late. I'm talking about **the pancreas**. Its islets of Langerhans contain two types of cells: alpha and beta cells. Their output controls the levels of your sugar in the blood. Alpha cells secrete glucagon which increases blood sugar levels, while beta cells secrete the *only* hormone that decreases your blood sugar: insulin. It does this by facilitating the absorption of surplus blood sugar by the rest of your cells.

When it comes to fuel regulation and appetite, the changes seen in normal aging resemble an intermediary stage of diabetes. The glucose is cleared out less efficiently. Besides, tissues develop insulin resistance. Metabolic syndrome is not only rampant in the elderly, but it accelerates aging in younger persons too.

The fasting glucose increases with 6-14 mg/dl per decade after 50 years old [78]. Just like with most other substances that get cleared less efficiently from the body, a higher level of glucose remains in the blood because the average tissues are less sensitized to the insulin hormone.

One of the criteria of determining senescence in humans and all other species is **a decrease in fertility**, whether this takes place almost abruptly like in female mammals with the onset of menopause or gradually like in male mammals.

The gonads produce the sex hormones corresponding to each gender, giving you secondary sex characteristics during the adult stage.

The ovaries produce progesterone and estrogens. Progesterone is most active during ovulation and pregnancy. During puberty, the estrogen release plays a major role in the development of breasts, womb, bones and the typical female adult body hair pattern.

The testes produce testosterone in males after puberty, causing the growth of their muscle mass and bones as well as the typical male adult body hair pattern. As boys become middle-aged men, testosterone gene variants may trigger androgenic alopecia or the typical male baldness seen in some elderly.

Needless to say, human aging is correlated with a decrease in your sex hormones no matter your gender. It starts almost abruptly in women with menopause taking place between 45 and 55 years old. This change happens in all women because of developmental limited oocyte stock.

During menopause, the ovarian follicles no longer respond to the upper command from the hypothalamus, gonadotropins or GnRH. Consequently, estrogen levels decrease [78].
At the same time, ovaries secrete less estradiol and testosterone around the age of 45-55 years old. Consequently,

the pituitary gland responds with an increased secretion of FSH and LH, a secretion which is gradually decreasing after the age of 75 years old.

The fertility decline starts gradually in men. The decrease in testosterone takes place gradually after the age of 20 with about 1%/year in total testosterone and 2%/year in free testosterone. The decrease in free compared to total testosterone takes place because the protein that binds testosterone – SHGB=sex-hormone-binding globulin – is increased with age. The former decrease in free testosterone is the andropause [78].

Hormonal treatment with sex hormones is associated with an increase in cardiovascular risk and in sex-hormone-dependent cancers, despite the minimum increase in bone mass and lean body mass. Apart from treating specific clinical symptoms in menopause for short-term, such treatments are not recommended [78].

In both men and women DHEA and DHEA sulfate decrease with age. This decrease takes place at a slower rate in healthier individuals with longer lifespans. Apart from being precursors to active androgens and estrogens, not much is known about their function and no changes are noticed after supplementation with these hormones. DHEA and DHEA-s may be markers of biological aging more than aging being a case of DHEA deficiency. Nevertheless, DHEA is sold as a food supplement [78].

Until now I presented you the major hormones secreted by glands, organs which specialized in secreting such important substances. Yet if I would stop here, I wouldn't tell you the whole story. *Many other hormones are produced locally in organs which don't belong to the endocrine system.*

In response to high blood pressure levels, your heart will secrete **the atrial natriuretic peptide** (ANP) hormone in order to bring them down to normal levels. It accomplishes this by dilating blood vessels as well as eliminating water and salt by your kidneys.

In response to low levels of oxygen in the blood, the kidneys immediately do a fine job of secreting **erythropoietin** (EPO), a substance which stimulates the bone marrow to produce more red blood cells. These cells will increase the capacity of the blood to carry oxygen to the hungry tissues.

The digestive system keeps the pace by secreting three local hormones: **cholecystokinin** (CCK), **secretin** and **gastrin**. In response to the food from your stomach, such hormones control the secretion of pancreatic juice, bile and gastric juice.

Few people know that the adipose tissue produces hormones on its own, including low levels of estrogens. It secretes leptin too, regulating appetite and energy usage by the body. The more fat you have, the more leptin you'll secrete, effectively communicating to your brain that you're not starving and that it shouldn't attempt to conserve energy or increase food intake.

The placenta of pregnant women secretes progesterone and human chorionic gonadotropin (HCG).

A series of hormones like prostaglandins and leukotrienes affect neighboring cells only. Prostaglandins increase local inflammation, while leukotrienes help heal the damage after prostaglandins did their job by reducing inflammation, so that the white blood cells get to clean the area of pathogens and damaged tissues.

The hormonal theory of aging has always been popular with people. Plenty of humans have been searching for age-related hormone variations as causes of aging and have consequently supported endocrine treatments as the "fountain of youth". Yet the hormonal system certainly plays an important part as one of the pacemakers of senescence.

Several changes happen in the human body as decades pass by.

First of all, the clearance of many substances decreases. Chemicals are eliminated at a slower and less efficient rate as decades pass by and for this reason harmless drugs in the young may pose serious problems in the elderly.

Secondly, many tissues develop resistance towards the action of hormones. This may be due to epigenetic changes in the receptors of these hormones. Consequently, certain types of hormone levels increase. This action may compensate the lack of tissue responsiveness.

Thirdly, many glands, specialized organs in the secretion of hormones, **get atrophied with age**. Consequently, despite an initial hypersecretion of hormones, their rate decreases towards the end of lifespan.

It is important to distinguish *age-related* from *disease-related* endocrine changes. This is easy in theory but difficult in practice. If a certain endocrine change is found in most "healthy" older people, that change is considered an age-related one. On the other hand, if the endocrine change is found in a minority of older persons, especially if they display clinical symptoms, then it is safe to tag it as disease-related [78].

Case Study: Low Hormone Levels in Long-lived Rodents

As you may have noticed from the lines above, many hormone levels decrease with age in human beings. The hormone replacement therapy was the next logical step in prolonging human youth. But here's the caveat: most aging research was and is done on short-lived species. Humans live for more than 5 times what is expected given our size [55]. It's the same for the naked mole rat [76].

As you'll see during this case study, many long-lived rodents and bats are characterized by low levels of several hormones including thyroid, insulin, glucocorticoid, vitamin D and sex hormones. All these hormone values may be the consequence of a deficiency in pituitary hormones, especially growth hormone. Calorie restriction produces similar results.

The common denominator of all these hormonal differences in long-lived mammals compared to shorter-lived ones may be *the Klotho longevity gene*. When overexpressed, this gene prolonged the lifespan of lab mice with 30% [67]. When mutated, the gene seemed to accelerate aging [66]. Klotho mutations are also associated with ectopic soft tissue calcification.

Low levels of the active form of vitamin D were noticed in [15]:

- naked mole rats
- nocturnal frugivorous and insectivorous bats

These animals neither get enough exposure to the sun, nor do they eat a high animal fat diet.

Such long-lived rodents and bats have the following glucose metabolism characteristics:

- low fasting glucose levels

- impaired glucose tolerance at supraphysiological loads

- low insulin levels

- extreme insulin sensitivity when this hormone is available

Even more, glycated levels in naked mole rats are low and constant with age [117]. Huge variations of glucose after meals in fruit eating bats may be due to their passive paracellular transport of glucose [20]. As a side note, these effects are noticed in the CRON (calorie restriction with optimal nutrition) community as well and impaired glucose tolerance which can progress to 'starvation diabetes' [9] is often a subject of worry.

When it comes to rodents and bats, there are seasonal changes in hormones regulating sugar: glucagon and insulin. During hibernation, glucagon may be increased to maintain fasting glucose levels, while insulin may be high to inhibit reproductive hormones [15].

Glucocorticoids are hormones animals employ to respond to acute stressors as well as to postpone reproduction until favorable conditions emerge. There is a glucocorticoid peak in rodents and bats during the breeding season.

Mammals have two main glucocorticoids: cortisol and corticosterone. There is a marked difference in the proportions of these hormones as follows [15]:

- corticosterone is the main glucocorticoid in short-lived rats and mice

- cortisol is the main glucocorticoid in larger mammals, bats and some long-lived rodents such as ground squirrels and rats

These two types of glucocorticoids may respond differently to the same acute stressor. Syrian hamsters maintain constant glucocorticoid levels, yet proportions change with age and the adrenal response is diminished. Cortisol increases, while corticosterone decreases with age [85]. Chronically calorie restricted rodents have elevated levels of corticosterone.

On the other hand, an increase of about 50% in glucocorticoid levels is experienced in accelerated aging species just after breeding:

- the *Oncorhynchus* Pacific salmon
- the *Antechinus stuartii* marsupial male mouse

In both cases, the glucocorticoid peak at the end of the breeding season may fatally suppress their immune system [34].

Long-lived squirrels, deer mice, bats and mole rats maintain low levels of T4. Such low thyroid levels are often associated with low core body temperatures in these long-lived species, just like calorie restricted rodents. The naked mole rat has one of the lowest levels of thyroid hormones, especially when compared to mice [17].

Thyroid hormones modulate the metabolic rate, hence they were often used in the lab to extend lifespan or accelerate aging. Their levels vary with:

- season
- food availability
- subterranean versus above ground animals

- hibernation

Thyroid hormones are responsible for the astonishing metamorphosis of tadpoles into frogs. Transcription – leading to turning genes on and off – is highly dependent on thyroid hormones.

Metamorphosis is different in amphibians compared to insects. Amphibians remodel their existing body tissues with the help of thyroid hormones, which directly control apoptosis or programmed cell death. On the other hand, insects destroy their larval tissues and build replacement cells [44].

The timing of metamorphosis in amphibians is influenced by the sensitivity of tissues to T4 and T3. As you'll see in the next pages on neoteny, thyroid hormone resistance plays a key role in amphibian species foregoing metamorphosis.

Most vertebrates experience reproductive decline with age. Species must produce not only the maximum number of offspring, but also the maximum number of *viable* offspring. Hence just like in plants whose seeds are capable of dormancy, many species synchronize their breeding with favorable times by using several hormones including insulin, glucocorticoids, sex steroids, T4, leptin and melatonin.

Animals receive environmental feedback from:

- the amount of daily light
- their nutritional status

And they use this information to schedule ovulation, mating, even pregnancy like it happens in bats [62].

Bats and long-lived rodents maintain *low reproductive hormone levels*. Concurrent with their slow growth, gestation periods are longer too. Naked mole rats and mice have almost the same size, yet naked mole rats take 4 times as long to reach reproductive maturity and their gestation period is almost 4 times as long: 77 days versus 19-20 days in mice. Gestation periods are even longer in bats: 120 days [15].

The dominant female naked mole rat shows no sign of reproductive senescence [16].

As a summary from this case study, long-lived rodents and bats are characterized by:

- low hormone levels
- high hormone sensitivity

Thyroid, insulin, vitamin D and glucocorticoid hormones are important for reacting to acute stress and scheduling reproduction during favorable times. They have a huge influence on the composition of cell membranes with effects such as [15]:

- increased membrane permeability
- increased metabolic rate
- modulation of membrane proteins
- modulation of membrane receptors' sensitivity
- increased mitochondria genesis
- more frequent cell division

All these happen at an inhibited rate in long-lived species with low hormone levels.

Is Aging a Form of Dehydration?

Imagine a bouquet of roses withering in a vase. Do you notice the wrinkling of petals, their decrease in volume?

They have plenty of water in the vase, so why do they dry? Do they develop an inability to absorb water or to preserve their cellular water levels? Lacking roots is not the answer as rose flowers wither on the vine too. In their place a fruit full of seeds will carry the life essence over time and space.

Just like people, not all roses produce fruits yet they all wither in the end. Water is life, which is robbed from us in steps as we age. The death of an old person is not sensational. It's not sudden. We barely notice it. They first lose their fertility, then their passion for life, the ability to do productive work, their memory, and finally whatever is left of them.

Elders lose their thirst instinct too. Running a marathon is easier than convincing an 80-year-old patient to drink a bottle of water. They may sip from it just to please you and then they'll leave it on the side.

Internal and extracellular housekeeping is less efficient with age. Excess weight becomes a common sight, just like plaques inside the blood vessels. Blood becomes more viscous, so clots become a problem no matter their location in the body. Lipofuscin and many other aging pigments are deposited everywhere – and we recognize the common age spots [116].

Enough said. A glass of water, please.

The Immune Pacemaker of Senescence

If the previous chapter provided you some insights regarding rapid and gradual aging, the immune system is key in what sets many negligible senescence species apart: their complete regeneration. As organisms became more complex, the immune system kept up. But unfortunately, a trade took place for a more specific immune response against restricted regenerative abilities.

Innate Versus Adaptive Immunity

An important role of any living being is to preserve itself. Whenever an injury takes place, the survival of the host is put in jeopardy. Hence creatures – especially multicellular ones – developed specialized tissues to act as soldiers in order to preserve their integrity.

Bacteria have their own defenses too. If infected with viruses, they begin secreting toxic enzymes that destroy the foreign DNA. And as they specifically destroy the pathogen DNA, they avoid attacking their own blueprint by methylating its own DNA and marking it as 'self', hence they distinguish self from non-self [82]. Each time the same viral species inserts its sneaky DNA in the bacteria, the latter will act the same. In other words, their 'immune' response – if we can call it like that – is non-specific. *It has no memory.*

The immune system detects the injury and it plays a huge role in removing damaged cells and replacing them with new, healthy ones. The immune system orchestrates the whole process by:

- detecting the injury in the first place
- killing the pathogen
- removing dead cells
- repairing the whole thing

Just like doctors around the world, the immune system will 'make do' with what it has at its disposal. In the case of adult humans, the immune system will finish its job leaving **scars** behind. Scars are deposits of fibrous tissue in what should have been tissues with the same form and function as the ones that were damaged. Such fibrous tissue is created by overexpressing collagen, albeit a weaker form of it. Fibrous tissue has a lower functional quality than the replaced tissue.

Apart from minor injuries, adult humans will repair damage by leaving scars behind. Developmentally speaking, that hasn't always been the case. Doing surgery on fetuses leaves no scars behind [18]. During those early times a human being is able to undergo complete regeneration with no fibrous tissue left behind. It does so under certain limits of course – human fetuses can still succumb to accidents and diseases.

When it comes to complete regeneration, tissue form and function are adequately reestablished after injury.

Minimalism pays off in the case of tissue regeneration – a characteristic of many negligible senescence species. **The more complex an organism is, the more limited its regenerative abilities are.** This may have something to do with *the complexity of the immune system.*

All organisms must solve the following problem: they must distinguish self from non-self. This is necessary when feeding. It's also a must in those organisms undergoing sexual reproduction.

Innate immunity is the simplest solution to this problem: fast and efficient. Its only downsize is non-specificity. If the organism encounters the same pathogen in the future, it will start all over again from scratch. Innate immunity is common in fungi, plants, unicellular animals and invertebrates [13].

When removing macrophage cells which are critical to the innate immune system, salamanders were unable to undergo limb regeneration and formed scar tissue instead [45].

Complex organisms such as vertebrates – starting with fish, amphibians, reptiles, birds and all the way to mammals – learned to *recognize* previously encountered pathogens. In other words, they developed **immunological memory**. It takes time to develop immunological experience just like it takes time to gather neural memories, but once this information is stored, it can help the organism better adapt to the environment. Although slow to develop, the immune system will attack much faster and stronger once it encounters the previous enemy.

But all this complexity has a downside. Just like big corporations get cluttered with bureaucracy – important changes taking an eternity to implement – so do complex organisms face limited regenerative abilities. Lacking a better solution, the immune system 'makes do' with what it has available and stitches the injury with fibrosis instead of normal cells.

Attacking takes energy and the organism's energy resources are limited. Hence elimination must always be weighted against tolerance. Sometimes it pays off to eliminate the intruders as otherwise the host's integrity is affected. Other times it is more efficient to spare that amount of energy and put it to better use by tolerating pathogens and even deriving benefits from them, like we do with our intestinal flora.

Lymphocytes are a subset of white blood cells and the main cells found in lymph, hence their name. B and T cells are the main effectors of adaptive immunity. Lymphocytes include natural killer cells too, but these are part of the innate immune system.

Adaptive, highly-targeted immunity acts through its specialized cells, B and T lymphocytes. Immature B cells are produced by stem cells in the bone marrow, hence their B name. T cells migrate to the thymus where they mature, hence their T name.

B and T lymphocytes are primitive in jawless fish like hagfish and lampreys. Yet they become more and more complex in cartilaginous and bony fish, amphibians, reptiles, birds and finally mammals [13].

Senescent Cells

When compared to youngsters, elderly people suffer from two things more frequently:

- they easily get infected, vaccines don't work so well on them and they get cancers
- they easily gain fat

Why?

The cells responsible for these functions become senescent and just like some old generals refuse to retire and do a bad job in the first place, senescent immune and adipose cells refuse to die [26].

You started life from one cell only. This cell divided exponentially up to a point. Afterwards surplus cells started committing programmed suicide or apoptosis. The end result is that you have the shape of a typical human being instead of being a ball of cells equally growing in all directions.

All is well, except that time passes by and the apoptosis process gets derailed. Two things happen.

Once specialized, most cells don't replicate anymore. Such cells commit apoptosis and the remaining tissue has to somehow compensate for their lack. It's like a factory where part of the workers kill themselves and the rest have to do the same amount of work as beforehand. Examples include the brain and the heart.

On the other hand, some cells *refuse* to commit suicide. Memory T cells once did a marvelous job, but once older, they refuse to make place for younger, naive ones. In consequence, elderly people are less equipped to deal with infections they never encountered before, are more prone to autoimmune diseases and often succumb to cancer.

As regards adipose tissue, we gain more visceral fat tissue as we count more candles on the birthday cake, while the subcutaneous fat remains all the same – and sometimes decreases.

The bad part about visceral fat is that it causes the following two age-related phenomena [26]:

- insulin resistance, possibly resistance towards other hormones as well
- pro-inflammatory signaling shift

Coming back to immune senescence, the thymus is a thoracic organ that has its developmental peak during adolescence. Thymic involution takes place in most vertebrates. Consequently, aging is characterized by *impaired adaptive cellular immunity*. Fewer naive T cells are produced and most T cells in the elderly are memory T cells.

Now what is the fate of the thymus in negligible senescence species?

Case Study: Thymic Involution in Negligible Senescence Species

Let's start with the immune system. You have a military budget made of two types of cells: T cells and B cells. You strive to maintain an equilibrium among them, while the budget remains *constant*. When young, most of your killer T cells are naive ones. They never destroyed foreign substances. As soon as they do, they attack the intruder – whether it's a foreign microorganism or a cell not recognized as 'self' anymore. Among the killer T cells responsible for destroying that specific type of non-self cell, you *keep* a couple of them just in case the intruder will enter the body again. If that happens, the immune attack will be much fiercer this time.

But as time goes on, more and more such 'generals' do their work and then refuse to retire! Not only this, they make life worse for the few T cells which are still naive – the humble soldiers. The military budget is constant, so the body will neither allocate more resources for new naive T cells, nor will it retire and destroy the non-functional 'generals' [26]. The thymus organ is a memory of the past.

A scientific paper linked thymus atrophy with the circulating level of sex hormones. Researchers were able to activate thymus regeneration in mice and elderly males by androgen blockade [107].

As I previously mentioned, evolutionarily we may have traded the ability to regenerate our organs against the gain of adaptive immunity.

Species that have extraordinary regenerative abilities include plants, colonial animals and a couple of free-living animals like the Hydra and planarian worms. **Plants employ innate immunity** only. **Invertebrates have neither thymus nor spleen** [58].

Species which are partially able to regenerate organs include the *Danio rerio* zebrafish and the *Ambystoma mexicanum* axolotl. The latter completes its thymic involution at the age of 30 years [2]. The zebrafish undergoes shrinkage of thymus as well [25]. These two species undergo thymic involution with age just like most other vertebrates.

Apart from the permanent thymic involution common with age in vertebrates, many animals undergo transient thymic involution during infections, pregnancy, malnutrition and hibernation [30]. Frogs, marmots and squirrels undergo transient thymic involution during winter [30].

79

The immune system is one of those pacemakers of senescence nobody seems to talk about, yet it is crucial in the body's attempt to repair injury. Simpler organisms are capable of reverse engineering their body when the inevitable damage happens – and this is the subject of the next chapter!

Reverse Engineering the Body

When it comes to life extension, regeneration research provides the best value for energy, time and money spent. **Regeneration is controlled growth.** It can be **complete**, when the new tissue is identical to the lost tissue in terms of structure and function. More frequently, regeneration is **incomplete**, when the new tissue shows a partial loss of structure and function. In this case, scars replace part of the lost tissue. Humans undergo complete regeneration in certain tissues like skin and the endometrium after menstruation. More commonly, they undergo incomplete regeneration with scar tissue formation after most injuries. So regeneration takes place in all organisms, but in different degrees.

Although largely regulated by asexual cellular processes, regeneration is not the same thing as vegetative reproduction. Yet the two are largely connected – it is more likely for complete regeneration to take place in organisms that reproduce asexually.

Cloning has been heralded as the new hope for regenerating damaged organs. Yet cloning as we know it produces older offspring compared to the ones conceived through sexual reproduction. This 'aging' effect is not seen in most species undergoing asexual reproduction, at least in those with immortal cellular clones. As we have previously seen, asexual reproduction *per se* is not the 'fountain of youth'. Once asymmetrical or uneven cell division takes place, unicellular organisms display signs of senescence too.

Regeneration takes place at different levels. In unfolding the story of regeneration, **cellular division** is a good place to begin understanding it.

Cellular division takes place in:

- single-celled organisms
- individual cells part of a multicellular organism

Single-celled organisms commonly include bacteria, fungi as well as protozoa or one-celled eukaryotes.

Their continuous division keeps the whole species alive. In this case, cell division can be:

- even or symmetric
- uneven or asymmetric

In the first case, the cell splits in two and all waste products are evenly divided among the two new cells. When it comes to uneven cell division, the mother cell keeps most waste products and after a couple of divisions, it will enter the path of senescence.

The common budding yeast *Saccharomyces cerevisiae* is one such organism where the mother cell shows signs of aging. Its size increases and its surface wrinkles as it loses its turgor. Compare that to another yeast, *Schizosaccharomyces pombe*, which divides evenly across generations [88].

Unlike bacteria and (single-celled) fungi, **protozoa** display both types of reproduction [88]:

- asexual reproduction where the cell contains large

polyploid macronuclei

- sexual reproduction where the cell contains small diploid micronuclei

Asexual reproduction in protozoa can be symmetrical or asymmetrical just like the rest of unicellular organisms. Examples of protozoa displaying asymmetric budding include *Acineta tuberosa* and *Tokophyra infusionum*. A couple of them manifest endogenous budding. I can't help but wonder whether that's where vertebrate mammal birth started from [88].

In those protozoa with two types of strains – one for each reproduction type – an interesting phenomenon unfolds. Such strains display clonal immortality when asexual and clonal aging when sexually reproducing [88].

Multicellular organisms come in many flavors. While single-celled fungi have either even or uneven cell division, multicellular fungi – or ***mushrooms*** as they are commonly called – are potentially immortal as they display ***indeterminate growth***. They continuously prolong their filamentous cells called hyphae and they form huge mycelium networks called thalli [84].

Fungi can grow to impressive sizes and lifespans. *Armillaria ostoyae* is the *largest* organism in the world with an estimated lifespan of more than 1,500 years old [19; 33; 88]. Yet potential immortality is not a characteristic of all multicellular fungi. A couple of them age after only 25 days [88].

Moving up the complexity scale, plants grow by cell division of meristems which are the equivalent of your stem cells. Meristems are usually stored in rhizomes, roots and shoots. You will often notice in common house plants like mint that parts of the plant are green and young while others dry up. As long as meristems still exist, the whole plant is alive.

Just like fungi, plants can form immense clonal organisms where individual plants form a single mass of underground roots or rhizomes. Such clonal organisms regularly reach lifespans of thousands of years.

But individual non-clonal plants like some trees can reach such lifespans on their own. The record breaker here is the Bristlecone pine [96].

At some point in history one-celled eukaryotes started cooperating to create the first metazoans or multicellular animals. Most basal metazoans are able to reorganize their bodies if you blend their cells. In other words, you can break them up into individual cells and they will reorganize into the whole animal afterwards. Isn't that fascinating?

Basal metazoans include 5 groups of animals [88]:

- *Myxozoa* which are small parasites
- Placozoa
- *Cnidaria* including corals, hydras and jellyfish
- *Ctenophora* or comb jellies
- *Porifera* or sponges

Not much is known about the *Myxozoa* parasites' lifespan and aging patterns [88].

Placozoa display differential aging patterns according to the type of reproduction a strain undergoes with senescence being the norm in those strains reproducing sexually.

Both *Porifera* and *Placozoa* have no definite nervous system. At the same time, these animals contain differentiated cells specializing in solving certain tasks.

Among the basal metazoans, cnidarians have an intermediate lifespan. They are able to reorganize, rejuvenate and regenerate their cells. Together with comb jellies, these groups of animals have a basic nervous system, muscles and simple sense organs.

Corals are colonies of individual polyps having a continuous cytoplasm in common [95]. Although individual polyps have short lifespans around a couple of years, most colonies survive for more than 4,000 years [95]. An example of coral aging is the colony of *Stylophora pistillata* which dies after a few years. Reproduction ceases around 6 months earlier [94].

A fascinating example of negligible senescence with *limited growth*, hydras contain two types of cells:

- non-dividing cells
- cells that divide continuously

The latter replace the former within 20 days, hence hydras maintain a steady adult size. Excess cells are turned into buds that will form new hydras farther away from the parent. Just like in sponges, their totipotent stem cells are able to migrate where healing is needed.

Perhaps the best known jellyfish is the *Turritopsis nutricula* 'immortal jelly' which is able to reverse its development stage [89].

Jellyfish are able to reproduce both sexually and asexually. They go through different life stages on their way to the next generation. Sexually mature jellyfish produce larvae which are mobile. They eventually settle and change from the previous planula stage to the actual polyp stage. The polyp grows and produces new jellyfish asexually. The new offspring further reproduce sexually [88].

In most jellyfish species, *sexually mature* jellyfish reproduce a couple of times and then die. But *Turritopsis nutricula* is different. When stressed, the adult jellyfish is able to turn back time to its previous asexual polyp stage [88].

Sponges are the animals with the longest lifespans [88]. Just like most metazoans, they are able to reproduce both sexually and asexually by forming gemmules, stolons or budding. **Sponges contain no separate germ cells**, as choanocytes and archaeocytes produce the necessary gametes for sexual reproduction [88].

Basal metazoans are animals with a simple anatomy. They are *the equivalent of protozoa colonies* and will often reorganize themselves if for any reason their cells get split up.

Bilateria are the next stage in complexity. As their name suggests, these animals exhibit bilateral symmetry. Their embryos produce the diversity of specialized cells starting from three cell layers: the endoderm, the mesoderm and the ectoderm.

Flatworms are the most primitive Bilateria [88]. Many individual flatworms may be potentially immortal, such as *Schmidtea polychroa* and *Schmidtea mediterranea*. Many of them are parasites and infect hosts, which are protected environments by definition.

Flatworms make use of three strategies to attain negligible senescence [88]:

- controlled shrinkage during starvation
- regeneration from body fragments
- asexual reproduction by fission

During starvation, flatworms are able to contract their bodies and shrink their size to the one common for juveniles, while keeping the prior proportion and distribution of cell types.

25% of planarian cells are neoblasts – somatic pluripotent stem cells. *Such cells are responsible for creating all the somatic cells of these worms.* An interesting experiment was done whereby irradiated worms were implanted with *one* neoblast each [114]. The regeneration was spectacular. Imagine what it would be like in humans to implant just one pluripotent stem cell and rejuvenate from the inside!

Here is another difference between us and them. Humans are animals in which the viability of germline stem cells is maintained by *telomere elongation* during embryogenesis. Asexual planarian strains need to maintain the viability of their somatic neoblasts, but they do this by *alternative splicing of telomerase*. They produce higher levels of active telomerase, thereby keeping the length of their telomeres constant even during fission reproduction and regeneration [108].

As you may have noticed, there is this trend of reduced incidence of asexual reproduction as complexity increases. In other words, species can have one of the following **types of rejuvenation during the adult stage** [88]:

- continuous rejuvenation
- limited rejuvenation
- restricted rejuvenation

Species with *continuous rejuvenation* are potentially immortal. Most simple metazoans are included here. Species with *limited rejuvenation* during their adult stage enjoy a moderately long lifespan and include more complex metazoans and a couple of vertebrates. Species with *restricted rejuvenation* like humans limit their regeneration ability to the embryo stage.

This classification goes hand in hand with the percent of totipotent primordial stem cells remaining in the adult animal [102]:

- species with *unlimited primordial stem cells* during the adult stage such as the archaeocytes in sponges, the I-cells in most Cnidaria and the neoblasts in flatworms. These species reproduce asexually and are potentially immortal.

- species with *restricted primordial stem cells* such as some annelids and mollusks. Both primordial germ cells and somatic cells continue into the adult and the animal enjoys limited pluripotency.

- species with *rudimentary primordial stem cells* like humans, where the primordial germ cells continue into the adult, but there are no adult pluripotent somatic cells anymore.

Case Study: Why Are Sponges Potentially Immortal?

Who knew that creatures traditionally used by people as natural bath sponges could be the longest-living animals known?

One step further from a colony of unicellular animals, an individual sponge is a multicellular and largely immobile animal. These fascinating organisms adopted all sorts of interesting strategies in the fight for survival and they seem to have made it.

Most sponges feed by filtering food particles. Hence they live in quiet, undisturbed waters. Since *water flow* is so important to extracting food and oxygen, they evolved something we could only dream of: **sponges are able to remold themselves**. In other words, they can change their body shape according to the environment. This card allows them to mold according to their substrate, encrusting rocks and other hard surfaces such as shells and corals [88].

Sponges are able to change their body shape by employing two mechanisms:

- The cells making up their external layers such as pinacocytes and choanocytes are not bound tightly like epithelial cells in your skin and mucous membranes.

- Their endoskeleton called the mesophyl can be continuously remodeled by specialized lophocyte cells.

All this can be summed up as living on unstable ground. Yet **flexibility pays off**.

There are two features potentially immortal species have and sponges make no exception. They reproduce asexually and their adult somatic cells are pluripotent.

Let's take them one by one.

Sponges reproduce sexually. Most of them are hermaphrodites, the same individual producing sperm and eggs, without having true gonads. Yet they preserved the ability to reproduce asexually and they do this in three ways:

- by fragmentation

- by budding

- by producing gemmules

Not all sponge fragments are able to recreate another individual from scratch. In order to do that, the fragment must contain at least two types of cells: collencytes which produce the mesohyl (their endoskeleton) and archaeocytes from which all other types of cells are derived [98].

When times are stressful, usually when temperature drops, sponges – and many other species – degenerate into gemmules. These tiny 'survival pods' stay dormant until better times come and the little creatures regenerate. Often such gemmules are retained within the parent sponge.

Asexual reproduction is not enough. Many species reproduce asexually without displaying extraordinary longevity. **The three Rs come next: reorganization, rejuvenation, regeneration.** Sponges have a huge advantage here: they are simple creatures. This time **minimalism pays off**. Sponges have no true tissues. Depending on the species, they have between 5-10 cell types [38]. They have no body symmetry. A symbol of flexibility, dare I say. *Most sponge cells are able to move around the body.* A few of them are able to *dedifferentiate* – transforming themselves from one type of differentiated cell into another, by-passing the usual stem cell route. There is another interesting ability in some sponge species: if you take one such animal, blend its contents and put those cells into water, they will reorganize themselves into a new sponge [88]. How cool is that?

I previously mentioned that one fragment must contain archaeocytes in order to recreate a sponge individual. Archaeocytes are totipotent cells. They can differentiate into any type of cell the sponge needs. This ability is totally lost in species like humans which are only able to express totipotency as embryos.

Sponges lack a complex immune system like you do, yet they are able to reject grafts from other species. What is fascinating though is that sponges will accept grafts from members of their own species [39].

I mentioned **sponges being potentially immortal**. The word 'potentially' is important here, because accidents are a part of life. They are predated upon by echinoderms, turtles and some fish. That is when they aren't turned into commodities such as natural bath sponges for human use.

But sponges understand **the sharing economy** better than anyone. They **collaborate** freely with anyone that will give them an advantage in survival.

For example, they'll often team up with photosynthesizing organisms like green algae, cyanobacteria and dinoflagellates. What's more, glass sponges – one of the three main classes of sponges, the other two being calcareous sponges and demosponges – have silica spicules which conduct light into the mesohyl, the endoskeleton where the green algae live in symbiosis with the sponge [12]. The sponge provides safety with plenty of light and the green algae provide some oxygen and organic matter as food. That looks like fair trade to me!

Sponges will often host shrimps too. Each type of shrimp belonging to the *Synalpheus* genus will inhabit a different sponge species enjoying not only safety from its host, but the larger food particles the latter can't digest [31].

Sponges are the animals with the maximum known lifespan [88]. They lack any protective shell. Sponges are immobile, lacking any means of escape. But they are flexible and evolved to synthesize a variety of unusual substances which pave the way to better drugs for humans.

Their lifespan varies wildly. Some sponge species survive for only a couple of years. At the same time, some demosponges grow their spicules very slowly at a rate of 0.2mm/year [98]. These species archive their chronological age in their spicules [59]. If we suppose the growth rate is constant, then a sponge with a diameter of 1 m could have at least 5,000 years. Another Antarctic specimen is estimated to be around 15,000 years old according to its growth curve [41; 54].

Sponges are fascinating organisms that can be easily grown at home, not to mention in a lab. Why aren't they used as a biological model of aging or rather their lack of aging?

Modular Growth and Aging

A feature often encountered in species with extraordinary regenerative abilities as adults is their modular structure. Such individuals undergo *modular growth*, so they are able – under certain limits- to regenerate parts of their bodies.

Their anatomy consists of modules which are attached together but if separated, each unit can survive on its own and even multiply. There is *no* distinction between somatic and germ cells. Yet a modular growth individual is not the same thing as a colony! Curious about examples of such species? Then read the following case study.

Case Study: Youth Is Forever Gone. Unless You Are a Hydra. Or an Immortal Jellyfish.

The hydra derives its name from its astonishing regenerative abilities, being able to continuously grow new individuals if you cut it in half again and again just like during that day when it was first studied by Swiss naturalist Abraham Trembley. An interesting fact about hydra is that its potential immortality can be wasted by knocking out one single gene: FoxO, a known tumor suppressor [11].

The *Turritopsis nutricula* jellyfish has three life stages:

- the planula larva with an expected lifespan of hours to days
- the postlarval polyp stage with an expected lifespan of days to years
- the adult or medusa stage with an expected lifespan of days to years

The first two stages are cystic ones.

This type of jellyfish evolved reverse development in order to increase its chances of self-preservation during harsh times. Such reverse development or ontogeny reversal is achieved by three related mechanisms [99]:

- transdifferentiation
- apoptosis or programmed cell death
- proliferation of interstitial cells

What this jellyfish does by 'ontogeny reversal' is a form of cloning.

The *Turritopsis nutricula* jellyfish is a biological model to study *transdifferentiation* in action. In other words, its cells managed to cut out the stem cell middleman being able to switch from one specialized type into another. It is not yet known whether all of its cells are able to accomplish this feat or only some of them. *Turritopsis nutricula* is an interesting biological model that could change the field of regenerative medicine for the better and remove the ethical complaints usually associated with it.

As regenerative abilities were lost with an increase in complexity, many genes switched their expression by being turned on, turned off, deleted or modified in any way. Comparative gene expression between closely-related species with regenerative ability differences could pave the way for huge progresses in regenerative medicine.

Some people maintain a fresh perspective on life, refusing to get bitter with age. The next chapter is about species that literally do the same.

Down The Neoteny Lane

Since mortality is high in most natural environments and predators are easy to come by, species have adopted the strategy of producing as many offspring as possible during the shortest amount of time in order to propagate their genes. This works great when there is an abundance of fuel and oxygen to burn it.

Yet other species have adopted the longevity strategy, where living longer *per se* guarantees the maximum number of offspring to propagate their genes. The environments inhabited by the latter are more hostile in terms of climate, but safer in terms of predators. Neoteny kicked in. While mostly derived from the same group of individuals living in affluent environments, harsh environmental factors influence their gene expression and prolong their juvenile phase without impending on their sexual reproductive maturation.

The harsh environmental conditions forcing such species to limit their growth and develop neoteny may include [91]:

- hypothermia given by altitude or latitude
- intermittent oxygen restriction
- intermittent calorie restriction
- climatic change

Neoteny is an interesting phenomenon and one of the common denominators of many negligible senescence species such as:

- the *Proteus anguinus* olm

- the *Ambystoma mexicanum* Mexican salamander

- the *Heterocephalus glaber* naked mole rat

- the *Balaena mysticetus* bowhead whale

These species are the focus of the next two case studies of neoteny in amphibians and mammals.

Case Study: Neoteny in Amphibians

Amphibians are creatures adapted to land and sea. Their life cycle has three main stages:

- egg

- larva

- adult

Metamorphosis is the rite of passage for the larva to become a full-blown adult. Salamanders, frogs and newts commonly undergo metamorphosis, an exquisite phenomenon controlled by the *thyroid* hormones to activate it and by the *prolactin* hormone to inhibit it [44].

A small number of amphibians have evolved *thyroid hormone resistance* and with it they dropped out of the metamorphosis program. They reach sexual maturity before completing their somatic growth. It is common for neotenic amphibians to live in harsh environments like dark underground caves in the case of the olm.

Neotenic urodeles – one of the main groups of amphibians including salamanders and newts – are divided in three categories according to the type of metamorphosis [15]:

- permanent or obligate – e.g. the *Proteus anguinus* olm, the *Necturus* American mudpuppy

- inducible – e.g. the *Ambystoma mexicanum* axolotl
- facultative – e.g. the *Ambystoma tigrinum* tiger salamander

The olm lives around 100 years, which is an eternity compared to the usual amphibian undergoing metamorphosis with an average lifespan of 5 years [113; 61].

There are major differences in terms of their regenerative abilities too. Neotenic amphibians are able to regenerate entire appendages, while those undergoing metamorphosis are still able to regenerate part of their tissues, but in a more limited way.

The *Proteus anguinus* **olm**, a symbol of Slovenian heritage, is a small salamander spending its life in underwater caves lying on the European bank of the Adriatic Sea.

This is the only species member within the *Proteus* genus. Its average lifespan is about 60 years with many individuals expected to reach their 100th birthday [113]. This is extreme for an amphibian.

Like most long-lived species, the olm is extremely stress-resistant. Its surrounding underwater cave environment is mostly deplete of resources and this has trained it to endure long-term starvation, decreasing its metabolic rate and even consuming some of its tissues [53].

This cave salamander has a slender hydrodynamic body with four growth-stunted legs. By horizontally waving its body, the little animal is able to advance through the dark water –

searching for prey, sometimes possible mates and at the same time, being on the lookout for possible predators. Not only the size of its legs is stunted, but the number of fingers each leg has (three in the front ones and two in the back ones) is less than the average amphibian.

Like most animals living in the dark, its vision did not have the opportunity to be trained. Though they are still able to detect light, its ancient regressed eyes play a rather decorative role. So what is an olm to do? It largely developed its non-vision senses as its flattened long body got accessorized with many useful sensors: chemoreceptors, smell receptors and sound receptors. The last ones are useful in detecting underwater vibrations.

Its skin has a translucent shade of pink. By exposing the animal to light, it slowly turns into darker shades proving that its ability to produce melanin, the skin pigment, is still there. Olms breathe with the help of gills and some rudimentary lungs.

The olm enjoys a few types of prey like small crabs, snails and rarely insects. It occasionally fasts as life in underground caves can be lacking in food for months at a time.

Cave salamanders are mostly social animals, excepting males that become territorial after becoming sexually mature. Since fights are energetically demanding, male olms prefer to display their strength rather than using it for destroying opponents.

The olm is a fascinating animal, both for its extraordinary longevity and its negligible senescence. It is able to undergo severe starvation. This cave salamander is sensitive to pollution which can reach higher toxic levels underwater compared to surface waters. For this reason, the cave salamander is considered a vulnerable species as regards its conservation status.

The axolotl or the Mexican salamander is an amphibian which developed neoteny and dropped out of the metamorphosis program. Metamorphosis can be induced in them by administering iodine or thyroxine hormone [115]. The poikilothermic axolotl preserves many juvenile amphibian traits such as external gills and a caudal fin once it reaches adulthood. Their limbs are underdeveloped. Their vestigial teeth are barely visible, hence they mainly feed by suction. Healthy individuals have the instinct of ingesting little stones to aid digestion and prevent impaction. Axolotls have limited abilities to switch their pigmentation for better camouflage. They are kept as pets in aquariums with water temperatures between 12 and 20 degrees Celsius.

Axolotls reach sexual maturity without undergoing morphological metamorphosis. Just like its close relative – the tiger salamander – axolotls undergo some form of cryptic or biochemical only metamorphosis [115].

Neoteny seems to be a survival mechanism in all salamanders, having evolved in environments characterized by:

- mountain and hill waters
- few sources of nutrients
- low iodine levels

In these harsh environments, neoteny allowed salamanders to reproduce using fewer energy resources compared to the terrestrial individuals that underwent metamorphosis.

When it comes to regeneration abilities, the Mexican salamander is able to regrow whole appendages without forming scars. It can easily accept transplants from the same species. All these abilities make the axolotl a very useful biological model for the study of regeneration and aging [97].

As its name suggests, the Mexican salamander is native to Central Mexico. Because of pollution, the introduction of predatory fish in their native habitats and the continuous extension of human population, axolotls are threatened by extinction.

Case Study: Neoteny in Mammals

Just like all mammal neonates, you used to inhabit a totally different environment when you were an embryo. The atmosphere was hypoxic and hypercapnic. In other words, you spent your days in a welcoming womb, albeit an oxygen-poor and carbon dioxide-rich one just like all mammal wombs. You didn't suffer any brain or heart damage, because up to a certain degree you were hypoxia-tolerant. Even telomerase was turned on, allowing you to grow and differentiate your initial cell. You were poikilothermic. You didn't need to have an internal switch that regulated your temperature. Your mother did it for you.

Once you landed in the current world, it all changed. The ratio between oxygen and carbon dioxide shifted. The drop in carbon dioxide started the breathing centers in your brain. Little by little, you started regulating your temperature just like most mammals do. You switched telomerase off cell by cell in most of your tissues.

Your outer world totally changed. That's not the rule for all mammals. Some of them inhabit such oxygen-poor and carbon dioxide-rich environments *most of their adult lives*. When coupled with *intermittent calorie restriction*, such conditioning may set the trail to *neoteny*. Examples include the bowhead whale and the naked mole rat.

The bowhead whale is the longest-lived mammal surviving more than 200 years [91]. It is the only baleen whale that lives its whole life around Arctic waters where temperatures are regularly close to 0 degrees Celsius. Other baleen whales migrate to temperate and tropical waters to breed and rear its young. Not the bowhead whale.

It undergoes intermittent oxygen restriction by diving in deep waters. The bowhead whale is a slow-moving animal, but resting periods are interspersed with bouts of intensive swimming and deep diving. In synergy with this, its diet is feast or famine. The bowhead whale feasts during the summer on a fat- and protein-rich diet of zooplankton. The nutritional stores are going to be of great use during winter, when the whale undergoes fasting. The animal maintains similar blood glucose levels as humans (86.8 mg/dl) with an increased lipid profile (cholesterol 409.7 mg/dl, triglycerides 287.0 mg/dl) [91]. Like most negligible senescence species discussed here, it undergoes very slow growth after weaning and it maintains a lower core body temperature than expected: 33.6 degrees Celsius [43]. The bowhead whale lives in a harsh, albeit protected environmental niche. Its only predators are killer whales and humans.

The naked mole rat doesn't dive in deep waters. Its complex burrow system though is just as oxygen-poor and carbon dioxide-rich as the deep waters periodically encountered by the bowhead whale. The small animal lives in the desert in sealed underground burrows and gets all the water it needs from its limited diet.

These small animals have minimally developed lungs.

Their hemoglobin has high oxygen affinity, just like all fetal hemoglobins do. Their diet is limited by rainfall. Individuals in charge of burrowing tunnels to find food have a much easier job when the ground is wet. That is when they may be lucky to find carbohydrate-rich tubers, one of these sustaining a whole colony for some time. But during the dry season, it is next to impossible to continue burrowing, so individuals have no choice but to fast. They are able to stretch their food more by displaying coprophagy. Yes, they eat their own feces. Sounds disgusting, but they may supplement their diet by recycling their endosymbionts – intestinal bacteria manufacturing vitamins among many other substances.

Compared to rodents of similar size, naked mole rats have a lower metabolic rate, coupled with lower free T4 thyroid hormone levels and a lower core body temperature than expected: 33.1 degrees Celsius [17; 91]. Such an interesting phenotype couldn't have evolved in an unsafe environment. Indeed, their only predators are snakes.

It's All About Neoteny

Deprivation of fuel, oxygen and comfortable external temperatures may impede normal development in otherwise healthy species, hence adult individuals preserve many features common in their younger life stage and exhibit a neotenic phenotype. Such modified gene expression may play an important role in the longevity and rarity of cancer in neotenic species.

During their early stages, animals display adaptations mostly preserved by neotenic species such as:

- efficient absorption, transportation and use of oxygen

- better DNA repair and regeneration abilities
- increased allocation of stem cells for tissue repairs

Some of these features are regained in a limited manner by *preconditioning* as adults. *Intermittent oxygen training/therapy* is used in athletes or patients for reconditioning their tolerance to reoxygenation and reperfusion [91].

Actually exercise *per se* produces physiological hypoxia [91]. Intermittent hypoxia is preferable to constant hypoxia as oxidative damage occurs not because of a drop in oxygen levels but during reoxygenation. Intermittent oxygen restriction may favor the selection of efficient mitochondria as well as increase the rate of autophagy, being known that apoptosis starts in the mitochondria.

Calorie restriction with optimal nutrition is the most robust method of life extension tested in several different species. A variant of calorie restriction practiced by people all over the world is intermittent fasting. Whether this is more or less efficient compared to classical calorie restriction is still a subject of debate. Still, intermittent calorie restriction favored the selection of several neotenic species.

Does Aging Start When Growth Stops?

Growth is a positive change in size over a period of time. Humans stop growing when reaching maturity. Most species undergo a similar cap on growing when reaching a certain size. Hence an interesting hypothesis was proposed stating that senescence starts when growth stops [5]. Certainly, many plants and animals undergo indeterminate growth and increase their fertility with age. Size itself may bring a load of benefits in keeping predators at bay. At the same time, growth brings fewer and fewer benefits after a certain tipping point. The link between aging and growth is the focus of this chapter. Does aging start when growth stops?

A cell that *evenly* divides itself in two **dilutes** its original damage. Having a fresh supply of somatic cells derived from your own DNA would largely correct many diseases of old age.

You started life from one primordial cell only – the egg cell and a set of genetic blueprints. This primordial cell was created by mixing **three types of DNA** [68]:

- maternal nuclear DNA
- maternal mitochondrial DNA
- paternal nuclear DNA

If the cell is viable, it starts dividing until it reaches a certain size, when the following happen:

- some cells **continue dividing** until they reach their own 'stop' signals
- some cells remain undifferentiated or partially

differentiated as a stock of **stem cells** for the rest of your life

- some cells **stop dividing**

Like retired persons who refuse to learn a new field, such cells are **too differentiated to do something else** so they continue living. In the end, these cells will age and then die.

An organism can grow by increasing the size of its cells (hypertrophy), by increasing the number of its cells (hyperplasia) or a combination of both. It is common for growth to cease as organisms reach maturity. When it doesn't, that amount of excess energy must be somehow consumed [7]. When this doesn't happen, aberrant growth during aging takes place. This energy excess can be spent on [7]:

- body growth in sexually reproducing indeterminate growth species like crustaceans
- the formation of clones in vegetatively reproducing species like sponges

Organisms are open biological systems. They mainly use energy for their own maintenance. Whatever is in excess of that will be allocated to growth, reproduction or both. Unlike determinate growers, indeterminate growth species continually make this allocation as future fecundity depends on the current choice between growth and reproduction. Such a pattern evolved in environments where an increase in size improves survival, whether by escaping predators or securing mates. It may have evolved in environments with variable season length [52].

Indeterminate growth species allocate equally between growth and reproduction. Among these, perennial plants seasonally switch between growth only during early season and reproduction only during late season [52]. If growth takes place before reproduction and growth itself has survival benefits, indeterminate growth pays off. Otherwise, determinate growth is enough [52].

Sometimes species are forced by environmental influences to allocate more towards one or the other, as it happened in an experiment where the lack of snail shells restricted the growth of hermit crabs. In consequence, they allocated more energy towards reproduction [52].

The ratio between the weight of the gonads and the weight of the body is called the gonadosomatic index. This index mainly stays the same or increases with size, at least in fish [52].

Even indeterminate growth species decrease their growth rates with age. For example, lobsters show asymptotic growth [46].

Most animals and plants seasonally lose a certain percentage of their somatic tissues. There is a critical level below which indeterminate growth doesn't take place [52].

Sometimes the same species can switch strategies from indeterminate to determinate as in the case of the *Schmidtea mediterranea* planarian worm. The same species can maintain populations with different growth strategies as is the case of the tomato plant. Most heirloom tomato varieties grow and keep growing their vines. Yet modern varieties have determinate growth: they grow up to a certain height and then they stop.

When it comes to employing one of the two strategies, adult *Danio rerio* zebrafish develop growth hormone resistance or insensitivity while its close relative the *Danio aequipinnatus* giant danio continues to respond to this hormone as an adult and is an indeterminate grower [6].

Indeterminate growth species maintain the ability to grow long after the organism reached maturity. Indeterminate growth species include some reptiles, fish, mollusks, a couple of plants and mushrooms. Determinate growth species limit growth to the juvenile stage by secreting retardant hormones. These two strategies are not set in stone. Even within the same taxonomic unit, more primitive species exhibit indeterminate growth while more derived branches limit their growth when reaching maturity [51].

Case Study: Indeterminate Growth in Crustaceans

Decapod crustaceans – literally ten-footed arthropods – provide great inspiration for progressing the field of gerontology. Their longevity is little studied.

There are two types of decapods:

- determinate growth ones which are short-lived species often displaying signs of mechanical senescence with lost appendages
- indeterminate growth ones which are long-lived species that evolved several aging-proof mechanisms

The indeterminate growth decapods include many species of *lobsters*, *crabs* and *shrimps*. Here are some biological strategies they evolved that allowed them to reach longer lifespans [1]:

- regeneration of lost appendages allowing them to keep on feeding, escaping predators and mating if mechanical senescence sets in

- removal of cellular waste

- life-long stem cell activity

- isolation of diseased tissues through melanization and encapsulation

- given the telomerase enzyme such tissues secrete, their incidence of cancer is virtually unknown

All these effects are mediated by their indeterminate growth.

Many memes circulate online, including the one on lobsters being immortal. Sure, they grow all their lives. Unlike us and all the other mammals, they express telomerase in their adult tissues. But they stop molting in their final years and they develop shell disease. Don't all these resemble pneumonia in the old man?

Telomerase is not the fountain of youth. Life extension is not telomere extension. Lobsters – and shrimps, and crabs – secrete telomerase all their adult lives and they don't die of cancer or at least most of them do not. I take from this that telomerase is not the only mechanism to reach negligible aging, but one among many. Adult tissue regeneration would be impossible without telomere extension.

Many species display indeterminate growth. The problem comes when differentiating the effects of increasing in chronological *age* versus the effects of growing in *size*. Even if being able to regenerate your tissues and grow indefinitely may seem like the fountain of youth, size *per se* imposes limits on these animals. Such limits may be falsely considered as signs of aging.

Arthropods face several limits to indeterminate growth:

- when compared to endoskeletons like bones, exoskeletons are heavy. Any added weight takes energy to keep it alive and use. At some point, the benefit of added growth is lower than the risk of immobilization and starvation.

- unlike large endoskeletons, a bigger exoskeleton covers the whole animal

- species undergoing indeterminate growth regularly *molt* in order to get rid of the smaller exoskeleton. Molting is energetically costly. Besides, the animal is very vulnerable during this process and the bigger the animal, the more difficult it is to hide during that time.

Apart from the aforementioned limits, indeterminate growth species can still exhibit signs of gradual aging. Examples include one beloved aquarium fish: the guppy.

The *Poecilia reticulata* guppy fish employs different growth strategies according to its gender [3]:

- male guppies are determinate growers

- female guppies are indeterminate growers

Yet guppy fish of both genders have an increased mortality rate with age and show reproductive decline too. In other words, they gradually age. Despite lacking kin networks and engaging in maternal care, female guppies have a post-reproductive lifespan of around one third of their total lifespan [92]. This is very similar to the grandmother effect in humans.

Danio rerio or the common zebrafish enjoyed by thousands of aquarium hobbyists around the world is an interesting biological model when it comes to studying aging.

This fish is able to do what patients in high-tech hospitals can only dream of: regenerate its heart despite its *limited* growth pattern once it reaches maturity. Its adult tissues express telomerase [74]. Perhaps that's how lipofuscin could not be detected at all [64]. Yet despite having all the cards in its favor, the zebrafish gradually ages nevertheless by accumulating more oxidized protein products as time passes by [64].

The Rate of Growth

Species differ not only in terms of using different growth strategies, but in their rate of growth too. In determinate growth species, this is highly dependent on food supplies through:

- physiological constraints
- cell repair mechanisms

This is similar to indeterminate growers during their juvenile phase. When they reach adulthood, competition between development and reproduction sets the rate of growth [65]. Reproduction is highly dependent on external mortality rates. Indeterminate growth is especially beneficial when mortality rate is high in small-sized organisms [73].

Most organisms grow at a submaximal, optimal rate. Since growth is highly dependent on food supplies, individuals always have to balance predation risk against foraging effort [29]. Slow growth leads to huge lifespan variations within the same species as is the case of the *Thuja occidentalis* white cedar with a lifespan ranging from 80 to 1653 years [69]. Slow growth pays off even in indeterminate growers such as the *Arctica islandica* clam.

Case Study: Aging in Bivalves

Hidden below sand on the sea floor lies an expert in survival. Time passes and this creature is just as good as new. It's called the ocean quahog or by its scientific name, *Arctica islandica*. Until recently, humankind had no idea of its longevity and most importantly, its negligible senescence.

Bivalve shells as a group archive information about their chronological age in annual forming age rings – at least when it comes to polar and temperate species. These tiny mollusks grow their shells two ways. One method is to add concentric rings around the horizontal borders of its valves. A second method is to increase the thickness of the shell itself. The latter is considered more precise in determining a bivalve's age. There are two types of bivalves: **epifaunal bivalves** that swim above the sediment and **infaunal bivalves** that commonly burrow in sediment.

The class of bivalves is a largely successful one full of survivors, especially the ones which burrow themselves under the sand, making them invisible from most predators. Those that do manage to eat their delicious flesh may either open their shell with their long beak – as is the case with gulls or other birds – or they may dig a hole in their shell and get the contents – like octopuses do. Rarely are they caught open.

The lifespan of bivalves ranges from 1-2 years in species like *Argopecten irradians, Pisidium spp., Donax spp.*, to more than 400 years in *Arctica islandica*. Unlike longer-lived bivalves of similar lifestyle, shorter-lived species undergo faster increase in oxidative damage as well as a rapid decrease in antioxidant capacity with age. Compared to other related bivalves, *Arctica islandica* enjoys increased antioxidant capacities as well as efficient and constant waste removal from its cells. The reason *Arctica islandica* is a fascinating biological model for the studying of aging is that the amount of lipofuscin and protein carbonyls is negligible, even in advanced age [1; 105].

The ocean quahog largely feeds on phytoplankton and diatoms by filtering water entering its shell. It breathes with the help of gills. These are its primary respiration surface. Its small heart has two auricles and a ventricle.

As regards locomotion, most bivalves lead a largely sedentary life – no gym nuts here. Apart from burrowing, they may use their only foot to move around the ground surface looking for a better place to feed. Some bivalves are able to swim, by rapidly opening and closing their valves.

Arctica islandica is the longest-living non-colonial animal with a maximum known lifespan of 407 years [105]. Like most negligible senescence species, it is extremely tolerant to anoxia. It has a protective shell and displays voluntary burrowing behavior during which the clam decreases its metabolism.

Unlike humans and even closely-related clams, *Arctica islandica* shifts the threshold at which it lives out of anaerobiosis. In other words, the ocean quahog enters anaerobiosis at lower oxygen partial pressures than expected.

Depending on the partial pressure of oxygen as measured in kPa, environments can be in a state of:

- normoxia at 21 kPa pO2 with the highest at sea level and further decreasing with altitude

- hypoxia at 2 kPa pO2

- anoxia at 0 kPa pO2

The ocean quahog clam depresses its metabolism when the partial pressure of oxygen decreases under 5 kPa, but it only shifts from aerobiosis to anaerobiosis at under 2 kPa [105]. The clam increases ventilation right before entering depressed metabolism. During this state, it postpones anaerobiosis until the pO2 decreases under 2 kPa. Postponing it until absolutely necessary for its survival is a smart strategy, as it avoids unnecessary buildup of acidic anaerobic metabolites. Because of this, the ocean quahog avoids their necessary removal which takes energy too [105].

In order to avoid anaerobic metabolite accumulation, the clam's mitochondria may use an alternative oxidase pathway through which they increase the rate of oxygen consumption while lowering tissue pO2 and the risk of free radical production [105].

Lipofuscin deposits slowly in its tissues, more in the gills respiratory organ, than in the mantle and adductor muscle [105]. Its heart cell turnover decreases by age 140, leading some to propose heart failure as the final limit on lifespan [71].

When young, Arctica islandica is a mahogany/golden brown clam. Later on, its shell stores more and more iron deposits and changes its color in black.

Long-lived mollusks divide their energy equally between their somatic and germ cell reserves. In *Arctica islandica* cell division is low, leading to an increase in cell diameter as the clam ages [105].

Unlike the short-lived and fast-growing *Aequipecten opercularis*, the ocean quahog's tissues grow extremely slowly and its oxygen consumption is low throughout most of its long lifespan [105].

Burrowing is an adaptive behavior for two reasons:

- it allowed such clams to avoid even more predators than expected by having only a hard shell.

- it allowed them to overwinter when food is scarce.

As expected from their maximum lifespan differences, Icelandic clams burrow more frequently than German clams and their burrowing interval lasts longer too [105].

Burrowing is not the only strategy used by these magnificent clams to decrease their metabolism. Active clams undergo respiratory breaks during which ventilation stops for *minutes*.

The ocean quahog tolerates temperatures of 0–16 degrees Celsius with an optimal interval of 6–10 degrees [105]. Different populations of *Arctica islandica* have different maximum expected lifespans. Clams living in the Icelandic waters may live more than 400 years, while those from Germany live around 150 years [105]. A large variation of temperature and salinity may be associated with a lower lifespan, while keeping the environment constant increases it. Oxidative stress increases with temperature. Yet antioxidants and stress proteins support, but do not cause longevity in *Arctica islandica* populations [105].

Consistent with its slow growth, the clam undergoes reproductive maturation at 10-14 years [105]. That's a lot of time to wait for such a small animal if it weren't for its predator avoidance strategies: a hard shell and burrowing.

Is Telomerase the New Fountain of Youth?

Mammals grow only during the embryonic and juvenile stages. They exhibit *determinate growth*. On the other hand, species with *indeterminate growth* often express telomerase in their somatic cells. They grow throughout their lives and often exhibit very slow senescence.

What could be the mechanism setting determinate and indeterminate growth apart? The telomerase enzyme is one such answer.

Telomeres contain repeated DNA sequences ending *linear* chromosomes. They are normally deleted with each cell division. Telomerase is an enzyme that synthesizes new telomeric repeat units so that the size of telomeres stays the same despite repeated cell divisions [35]. This enzyme is destroyed by heating.

The 'vertebrate' (TTAGGG)n telomeric repeat sequence is common in most multicellular organisms, including:

- humans
- invertebrate lower metazoans including sponges, corals, jellyfish, comb jellies and Placozoa
- Bilateria invertebrates like flat worms, velvet worms, most ringed worms, mollusks, echinoderms and tunicates
- vertebrates including fish, amphibians, reptiles, birds

  Exceptions include roundworms and arthropods [46].

The nematode telomere motif is (TTAGGC)n, while the arthropod telomere motif is (TTAGG)n. Beetles (order *Coleoptera*) lost the arthropod telomere motif and likely employ alternative lengthening of telomeres [46].

There are a set of important parameters to study in long-lived species:

- telomere length, usually measured in kb or kilo base pairs and how this varies among individuals of the same species and between different species
- variability of telomeres with age – whether they shorten, maintain or elongate with age
- telomeric DNA as a percentage of total DNA
- ratio between telomerase expression and ALT or alternative lengthening of telomeres as both are strategies to maintain or elongate telomeres during cell division
- when and where is telomerase expressed – which tissues and which cell cycle stages need the activation of telomerase
- how do all these influence the rate of senescence, the cancer rate and the average and maximum lifespan of species

Poikilotherms like invertebrates, fish, amphibians and reptiles persistently express telomerase in adult somatic tissues. This could have an impact on their regeneration abilities. Temperature increases metabolism, hence it may increase cancer mutation rates. This could be the reason for which endotherms like birds and mammals repress telomerase in their adult somatic tissues as a cancer-protection mechanism [46]. As previously mentioned, endotherms have higher metabolic rates than poikilotherms.

In sea urchins at least, long-lived species like *Strongylocentrotus franciscanus* and medium-lived ones like *Strongylocentrotus purpuratus* have short telomere lengths of around 5 kb, while short-lived species like *Lytechinus variegatus* have long telomeres of around 20 kb. Nevertheless, no telomere shortening takes place in all these three examples [36; 46]. Yet when it comes to birds, telomeric shortening takes place much faster in short-lived species compared to long-lived ones [46].

The length of human telomeres is 10–15 kb. Unlike humans, rodents have extremely long telomeres of 25–150 kb which don't decrease with age [46]. Rodents have a much higher cancer rate than humans which is not increased further in telomerase knockout mice [46]. Rodents usually have telomeres longer than 30 kb [46], while telomerase activity seems to inversely correlate with body mass. In other words, larger rodents express less telomerase. When it comes to rodents, there is no correlation between telomere length with size or lifespan [46].

The telomere length of the *Chrysemys picta* painted turtle is over 60 kb. Apparently, this length and their subsequent growth rate is maintained with age [46]. The related *Emys orbicularis* European freshwater turtle doesn't show any signs of senescence according to current knowledge, maintaining its 20 kb telomeres constant with age [46].

In line with activating telomerase, species like *Arctica islandica* maintain their telomere length during their lifespan [48].

When it comes to plants, these have two types of tissues:

- telomerase positive meristematic tissues forming roots, rhizomes and shoots

- telomerase negative non-dividing cells forming leaves and axillary buds

Telomerase is not expressed equally during the life cycle of an organism, in its tissues or during the cell cycle stage. Telomerase peaks in proliferative tissues and it is downregulated in postmitotic ones. Even in colonial animals like the *Botryllus schlosseri* golden star tunicate, telomerase peaks in bud rudiments and further decreases in its zooids. In other words, telomerase activity peaks in progenitor and stem cells and it is downregulated during differentiation [46]. Telomerase is highly expressed in cells which actively divide and it is downregulated during quiescence.

Consequently, telomerase expression seems to correlate with the regenerative potential of a species or at least that of its germ and stem cells and not so much with its maximum lifespan.

Case Study: Same Species, Different Telomerase Expression

The *Schmidtea mediterranea* planarian flatworm is a species having two types of strains: some reproduce sexually and others asexually.

The worm reproduces asexually by fission, keeping its telomeres constant. Just like humans, sexually reproducing planarians maintain constant telomeres during embryogenesis and in germ cells. Yet in both cases the length of their telomeres decreases with age [108].

Despite different telomerase expression, both strains of *Schmidtea mediterranea* enjoy an indefinite capacity to regenerate. Planarian adult stem cells are able to renew the animal's differentiated tissues, but this process doesn't depend on its telomerase activity or on the length of telomeres [108].

According to the telomere loss theory, telomere shortening leads to the aging of cells and that of the whole organism [10].

Telomerase Gene Therapy

If telomeres shorten with age and that leads to replicative senescence, doesn't it make sense to insert the telomerase enzyme in somatic cells? Apparently yes, but the greatest fear is that such a process may lead to the onset of cancer cells.

Here are two reasons for which this is an unfounded fear:

- telomerase is expressed in the embryonic cells and germ cells of several organisms undergoing senescence. Besides, it is expressed in the somatic cells of species with indeterminate growth and/or vegetative reproduction. Apparently, such cells and organisms do not undergo malignant transformation at a higher rate than expected.

- shorter telomeres lead to genomic instability which may be the main risk factor of developing cancer. A rarity in children, cancer runs rampant with age precisely when telomeres are shorter.

In line with the previous arguments, telomerase was inserted in adult and old mice with the help of a viral vector. A life extension of 24 % in the adults and 13 % in the elderly was achieved. Compared to controls, the treated mice did not develop cancer at a higher rate [27].

The next case study includes a group of species that express telomerase as adults. Their cancer rate is surprisingly low.

Case Study: Sea Urchins

Deep below sea level lies a beautiful and at the same time threatening animal. A red ball of spines ready to sting you if you dare to touch it. Upon closer inspection, you identify it as the red sea urchin, a long-lived creature which doesn't seem to age like the rest of us.

The animal displays five-point radial symmetry or pentamerism. Its size ranges between 6 and 36 cm. Its test, a threatening shell of spines, allows it to escape many predators that find it hard to reach its soft and carefully hidden inner parts.

The red sea urchin loves to feast on seaweed, but it won't refuse delicacies such as sea cucumbers, mussels, polychaetes, sponges, brittle stars and crinoids. The roe of sea urchins is a delicacy in many cuisines around the world, being especially prized in Japan. Apart from humans, their crown of spines will not protect them against sea otters, California sheephead fish and wolf eels.

The red sea urchin has a spherical body. Its inner organs are sheltered by the hard test made of calcium carbonate plates. Each plate is covered by a round tubercle over which spines are attached. It is able to move on the floor of the sea by using its tube feet, which are activated like a hydraulic device pumping water into and out of the feet.

The mouth of the red sea urchin lies in the center of its oral surface, surrounded by a softer lip-like tissue and five small and bony teeth jaws. Unlike mammals, their teeth are self-sharpening, so they do not dent with age.

Locomotion and senses are controlled by a large neural ring which encircles the mouth, with five main nerves radiating from it along the radial canals of the water vascular system. These main nerves further branch into finer instruments of control for the urchin's feet, spines and pedicellariae structures. The red sea urchin senses the outer bigger world through touch, light and chemicals.

In order to breathe, the red sea urchin makes use of two respiratory organs: gills and tube feet. Not only does it display a double gas exchange surface, the red sea urchin excels in circulation redundancy by having a water vascular system and a hemal system containing blood.

Sea urchins reproduce sexually by releasing eggs and sperm in the sea. Fertilization takes place outside the female's body. Some urchin species afford a greater degree of protection to the fertilized eggs by holding them among the female's spines.

Once fertilized, the egg undergoes a number of rapid cell divisions for 12 hours at the end of which it turns itself into a free-swimming blastula embryo. The next stage is a cone-shaped echinopluteus larva. Depending on the abundance of the inherited egg yolk and whether they need to feed themselves on something else, some of these larvae develop limbs while others do without them.

The larva needs a couple of months to complete its genetic instructions of growth and development. Afterwards it sinks to the bottom of the sea where it undergoes the adult metamorphosis stage, a point at which they forgo the initial larval bilateral symmetry in order to adopt the five-point radial symmetry.

Sea urchins are characterized by:
- indeterminate growth
- spine regeneration
- constant fertility rate with age

Yet species of sea urchins have various lifespans [36]:
- 3–4 years in the case of the *Lytechinus variegatus* sea urchin
- more than 50 years in the case of the *Strongylocentrotus purpuratus* purple sea urchin
- more than 100 years in the case of the *Strongylocentrotus franciscanus* red sea urchin

Telomerase is present in the early and adult stages of all these sea urchins and their telomere lengths show no age-related shortening. So telomere length is not the mechanism underlying their lifespan differences. Besides, there are very few cases of reported neoplasms in these animals [36].

Apart from the lack of telomere shortening with age or differences between germ and somatic cells, there is no difference in oxidative damage between sea urchins with different lifespans. All of them maintain regeneration abilities with age [10].

Perennial Plants and Their Regenerating Roots

Although negligible senescence species are encountered in the animal as well as in the plant kingdom, traditionally there has been a lack of interest in studying plants in the gerontology field. But are we wise for doing this? *Is there something plants can teach us that animals can't? Aren't they easier to study, both in terms of the necessary instruments and the ethical problems their study may presume?*

There are three kinds of plants: annual, biennial and perennial ones.

Annual plants undergo their whole life cycle from germination to the formation of the next set of seeds in less than one year. Examples include peas, lettuce, watermelon and many others.

Biennials take two years to complete their biological cycle, growing their vegetative structures in the first year and producing flowers and seeds in the second one. Afterwards the plant dies. Spinach is such an example.

On the other hand, perennial plants survive more than two years. How do they do this? They leave behind not only their seeds like the annual plants do, but part of their roots as well! They reproduce not only with the help of seeds, but by employing vegetative reproduction too. If you cut one such root in half and you plant each half in appropriate soils, two individual plants will be born. It is not only the root that can produce new individuals, but part of their underground stems as

124

well – and the two of them are called *rootstock*.

Perennial plants include:

- herbs – like alfalfa, red clover
- trees and shrubs – like the maple, the apple, the evergreen trees, the sequoia

Perennial plants have varying lifespans depending on whether they are:

- non-clonal
- clonal

The longevity record for a non-clonal (perennial) plant belongs to the *Pinus longaeva* Great Basin bristlecone pine at 5,062 years [96].

If life is a chess game where the king must be defended at all costs, then:

- the meristem (the plant tissue of undifferentiated cells) is the actual king
- the plant's growth modules are the chess boards
- the plant's vascular tissues, leaves and flowers are the chess functional pieces

You can see now why it pays to switch from an annual plant to a perennial one. Annual plants play short chess games, at the end of which all kings are dead (no more meristematic cells here!), while perennial plants risk more complexity for the sake of asynchronous chess games. Their games start and end in different parts of the plant at different times. They always have a supply of kings available so they are able to go through another winter [81].

When it comes to studying aging in plants, the three things to learn from these surviving plants are:

- meristematic cells are the plant's equivalent to our stem cells and it is in such cells that the key to long lifespans reside

- it pays to take it slowly. We can't but hope that suspended animation in humans will become as reliable as dormancy is in plants

- these plants survive through their ability to form new roots.

   As a far-fetched comparison to humans, what sets the optimist 90-year old – who survived two world wars and the death of several family members and friends – apart from the complaining 50-year old – with no serious diseases to battle – is the ability to form new roots: new friendships, new relationships, new hobbies, plans and hopes!

Case Study: The Bristlecone Pine

Negligible senescence species are found not only in deep waters, but living on the top of the mountains as well. The *Pinus longaeva* Great Basin bristlecone pine is one such example.

Its rich crown of needles has been a witness to the American State gaining its independence. It may have been witnessing the genocides committed by the Europeans as well. Yet this pine species – and the individual members of it – have continued to survive, apparently unaffected by our dismal decay.

One such individual pine is the oldest known living non-clonal organism celebrating 5062 years old [96]. That's a lot of life.

Living in harsh environments keeps them away from most predators. Fortunately, its most long-lived members are safe from humans as well, since their exact location in The White Mountains of California is virtually unknown.

One of its partners is the tiny *Nucifraga columbiana* Clark's nutcracker bird which hoards the pine seeds as a food resource. Thanks to its failing memory, some seeds grow into trees – at several distances away from the parent tree.

Individuals of bristlecone pine are either male or female. Each gender produces a specific type of cone. After a ripening of 16 months, the cones open and release the seeds immediately, which may be either carried away by the wind or rather being buried by the Clark's nutcracker.

Pinus longaeva has a protected status in many American areas like the Ancient Bristlecone Pine Forest in the White Mountains of California and the Great Basin National Park in Nevada.

Compared to trees which seasonally shed their leaves, evergreen ones have a slower rate of growth. They can afford to do that as they don't have to spend additional energy in recovering nutrients from dead leaves with the help of their roots. Yet what makes the bristlecone pine even more exceptional is that its leaves have the longest lifespan among all leaves, some reaching 45 years before falling to the ground [32].

Unitary Versus Colonial Organisms

Which is the border of an organism? Sometimes it is difficult to draw the line. You are a unitary organism because all your cells share *the same DNA*. Sure, your germ cells may contain any combination of only one chromosome of each pair, but they don't contain the genes of any other individual but yourself. You are bordered by your skin as well as by your mucous membranes. You are all one unitary organism.

But other organisms have their cellular units spread out. Borders are unclear. Such colonial organisms can be made of unicellular modules like bacteria or multicellular units like sea sponges or trees. A zooid is an individual part of a colonial organism. Zooids can either be directly connected by tissue as in the case of corals or share a common exoskeleton as in the case of moss animals. What they share together is *the same DNA*. Such colonies spread vegetatively.

With or without sexual reproduction, when it comes to massive longevity – meaning longevity counted not in days, months, years or decades, but in *millennia* – the record breakers are the colonies of organisms, where each organism is an *identical genetic copy* of all its neighbors.

Here are a couple of such record breakers [106]. One caveat: a couple of trees have wildly varying estimated lifespans [72]. Such trees have often evolved in harsh environmental niches [79]. As they aged, their trunks got hollow. Consequently, dating them exactly became difficult.

Colonial fungi:

- *Armillaria ostoyae* honey mushroom 2,400 years

128

Colonial lichens – symbiotic organisms composed of a fungus and photosynthetic cyanobacteria or a green alga:

- *Rhizocarpon geographicum* map lichens 3,000–5,000 years

Colonial bacteria:

- *Actinobacteria* Siberian bacteria 400,000–600,000 years – these are not in a state of suspended animation as they undergo DNA repair below freezing
- Stromatolites mainly rock bound *Cyanobacteria* 2,000–3,000 years

Unitary trees:

- *Pinus longaeva* bristlecone pine 5,062 years
- *Sequoiadendron giganteum* giant sequoia 2,150–2,890 years
- *Taxodium ascendens* Pond cypress 3,500 years
- *Taxus baccata* European yew 2,000–5,000 years
- *Castanea sativa* sweet chestnut 3,000 years
- *Olea europaea* olive 3,000 years
- *Cryptomerica japonica* Japanese cedar 2,180–7,000 years
- *Ficus religiosa* sacred fig 2,294+ years
- *Adansonia digitata* baobab 2,000 years
- *Juniperus occidentalis* Western juniper 2,675 years

Colonial trees – when a forest is actually one tree organism:

- *Populus tremuloides* Quaking aspen 80,000 years

- *Nothofagus moorei* Antarctic beech 6,000–12,000 years

- *Lagarostrobos franklinii* Huon pine 10,500 years where unitary stems are long-lived too surviving more than 2,000 years

- *Eucalyptus phylacis* Eucalyptus 13,000 years

- *Picea abies* Norway spruce 9,500 years

- *Quercus palmeri* Palmer's oak 13,000 years

- *Fitzroya cupressoides* Patagonian cypress 2,200 years

- the underground forests of *Parinari capensis* and other such species 13,000 years – in order to survive extended fires, these deciduous trees moved their trunks and roots underground, keeping only their crown above ground which is easily regenerated. Its central stem is sometimes referred to as a rhizome. Their age was estimated from their growth rates.

- *Welwitschia mirabilis* Welwitschia 2,000 years – this living fossil is the national plant of Namibia. The plant has only two leaves derived from its cotyledon first leaves. The plant forgot to grow up and stays stuck in its juvenile state. It undergoes indeterminate growth while living in the desert.

Colonial plants – other than trees:

- *Larrea tridentata* creosote bush 12,000 years – this plant can survive without water for two years because of its developed roots

- *Yucca schidigera* Mojave yucca

- *Gaylussacia brachycera* box huckleberry 8,000–13,000

years

- *Azorella compacta* Yareta 3,000 years – this ancient plant grows at a rate of less than 1cm/year in the Atacama Desert
- *Posidonia oceanica* Neptune grass 100,000 years
- *Lomatia tasmanica* Tasmanian lomatia 43,600 years
- *Chorisodontium aciphyllum* Antarctic moss 5,500 years
- *Polytrichum-Chorisodontium* Antarctic moss 2,200 years

Colonial animals:

- *Colpophyllia natans* brain coral 2,000 years
- *Leiopathes* coral 4,270 years
- the barrel or volcano sponge 15,000 years

Indeterminate growth can't take place forever. Beyond a certain tipping point, there are more risks than benefits to having a large size. At the same time, determinate growth may pose caps to the maximum lifespan of species. At the very least, determinate growers may actually continue growing in an aberrant way.

If youthful times are characterized by *exorbitant growth*, aging is a strange mix of *growth cessation* and *aberrant growth* at the same time. Macroscopically you get shorter and frailer with age. Microscopically you secrete less growth hormone with each decade.

According to the quasi-programmed aging theory, aging is characterized by [8]:

- hypertrophy of several organs – the heart being the classical example

- hyperplasia – with an increased risk of developing cancer
- hyperfunction – the body trying to compensate for all the changes it incurs.

The focus of the next chapter is the worst case of aberrant growth – cancer – a reality that increases in incidence with age in most species.

Cancer

What is the number one disease killing the elderly besides heart disease?

Cancer. A dreadful word that conjures all sorts of evil things, cancer runs rampant as age increases.

What do all cancers have in common?

They start from one (or several) nuclear mutation(s) [26]. In the case of non-inherited cancers, the disease starts from epimutations, which are DNA changes caused by the environment. Compared to mitochondrial mutations, nuclear ones take place at a far slower rate and most of the time they are inoffensive with one exception: cancer.

'The War on Cancer' was started by President Nixon in 1971 and although major progresses took place in its diagnosis and treatment, cancer still exists and most importantly, it still kills. What makes **cancer cells** so resistant to medical treatment is that they **undergo natural selection** and **you can't outsmart evolution**.

All cancer cells need:

- a serious supply of **glucose** – that is why obese and diabetic patients have an increased risk of cancer
- a good network of **blood vessels** which they produce *de novo* as they secrete many vascular growth factors
- telomere elongation mechanisms like **telomerase** to divide *forever*

Cancer cells are able to grow into tumors when they escape

immune surveillance and apoptosis.

Novel cancer drugs may act by decreasing the amount of available glucose, by blocking the vascular growth factors tumors secrete and finally by deactivating the telomerase enzyme.

According to the strategy of achieving immortality, cancer cells may [26]:

- turn on telomerase in 90% of the cases
- lengthen their telomeres through a lesser-known phenomenon called Alternative Lengthening of Telomeres (ALT) in 10% of the cases

Because cancer cells need telomerase to divide, shutting it down or deleting it altogether is a strategy that is bound to change the treatment of this disease. It is seductively simple in theory but it takes much effort, time and money to put it in practice.

The Paradox of Peto

The easiest way to be sure you'll never suffer from cancer is to be a unicellular organism. Does that mean cancer is an inescapable fact of life for long-lived and often complex beings?

Cancer starts out as a random mutation in one cell. Reason follows that the more cells an organism has, the higher its risk of developing cancer. Reality tells a different story. This paradox is named after Richard Peto, the British epidemiologist who first described it. Animals with a larger body mass rarely get cancer. Because they have few predators, they were pressured in evolving mechanisms for keeping cancer at bay. Smaller animals on the other hand didn't suffer the same pressure because of higher mortality risks. Bringing them in safer captive environments unveils the easiness with which their cells become malignant.

For this reason, many cancer-proofing mechanisms are waiting to be discovered in animals with a large body mass, species like elephants and whales. Needless to say, these are not the usual biological models of oncology research today.

Cancer cells are a product of evolution. They are extremely competitive against the normal cells of a host. What would stop them from being individualistic with one another too?

According to *the hypertumor hypothesis*, as tumors grow larger, younger malignant cells take over their parents and outstrip them of their blood supply, resulting in tumor ischemic necrosis. Metastasis could then be an adaptation of cancer cells to avoid competition for resources within a growing tumor [83].

Tumors can be lethal through their *size*, compressing vital organs. This lethality increases if the tumor is already metastasized.

But tumors can also be lethal because of their abundance of *blood vessels*, which can squeeze energy resources from normal tissues leading to starvation of the host. These blood vessels can often clot as cancer itself increases the blood's viscosity.

Researchers have launched all sorts of hypotheses for explaining the Peto's paradox.

There is an inverse relation between size and metabolism rate. Namely the bigger the animal, the slower its metabolic rate. Since increased metabolism drives tumorigenesis, it may be that slowing down metabolism with increased size is what reduces the risk of developing cancer in large animals [24].

At the same time, the cells of larger organisms are bigger in size and divide slowly. The energy turnover is minimal, hence they have a lower risk of developing genetic mutations [75].

Case Study: Cancer in Long-Lived Species

40% of all mammalian species are rodents [47]. They are ubiquitous. Their variety in terms of body mass and maximum lifespan makes them good biological models for studying not only aging, but cancer-proofing mechanisms as well. Although belonging to the same genetic family, rodents' maximum lifespan varies from 3 to more than 30 years. Naked mole rats are the longest-lived rodents known [14]. By comparison, calorie restriction with optimal nutrition increases maximum lifespan in mice from 3 to 5 years at most!

According to their body mass and maximum lifespan, rodents can belong to one of the following four groups [47]:

- large body mass with average lifespan rodents – the capybara
- large body mass long-lived rodents – the beaver, the porcupine
- small body mass short-lived rodents – the common mouse
- small body mass long-lived rodents – the naked mole rat, the blind mole rat, the gray squirrel, the woodchuck, the

chipmunk

Why is there so much variety in lifespan when rodents are so similar to each other in their genetic blueprint?

For one reason or another, some rodents escaped their initial dangerous environment. Porcupines developed spines. Beavers built logs. Naked and blind mole rats conquered the underground. Sometimes their sheer size eliminated the danger of being preyed upon. All these led to novel genetic strategies that postponed cancer and other age-related diseases, some of these still waiting to be discovered.

Cancer may conquer the whole body, but it always starts in one cell. There are two types of genes and mutations that will make the cell malignant:

- oncogenes
- tumor suppressor genes

The random mutation may either turn on an oncogene or turn off a tumor suppressor gene. In either case, the malignant cell starts a war largely unsuspected in the beginning.

Rodents with a larger body mass developed the same cancer prevention mechanism as humans: they silenced the telomerase gene in most somatic cells. This mechanism works on the short-term, as the resulting genomic instability may actually increase the risk of cancer. Switching off telomerase leads to replicative senescence where whole cellular lines die off. When these tissues can't replicate anymore, many signs of aging become visible. The weight at which such a mechanism usually gets in action – at least in rodents – is around 10 kg [47].

Smaller long-lived rodents developed cancer prevention mechanisms *despite* abundant secretion of telomerase in their somatic cells. Such rodents include the gray squirrel, the woodchuck and the chipmunk [47].

A larger body mass may be correlated with a longer lifespan when comparing one species with the other. Certainly elephants and whales lead longer lives compared to most mammals. But when it comes to individuals within the same species, larger ones will actually be short-lived [47]. Larger breeds of dogs live less and age faster than smaller ones [47]. Laron syndrome patients are characterized by dwarfism, but they are largely protected from age-related diseases such as cardiovascular disorders and cancer [47].

Normal cells arrest their growth as soon as they touch another cell. Logic follows that the more *sensitive* a cell is to contact inhibition, the less likely it is to become a rugged individualist malignant one. The cells of **the naked mole rat** *Heterocephalus glaber* can certainly be characterized this way. The extracellular matrix has a lot to do with it as cells secrete a special form of hyaluronan. This antioxidant substance makes the extracellular matrix more viscous than your own or the common mouse's one [47]:

- naked mole rat hyaluronan is five times larger than the one found in mice and humans

- the same type of hyaluronan has a lower turnover because of increased resistance to the enzyme that normally breaks it down

Another genetic strategy of the naked mole rat is their cleaved 28S ribosomal RNA [47]. In less fancy words, their protein translation from the genetic blueprint is accurate and stays accurate!

The *Spalax ehrenbergi* **blind mole rat** lives underground. It developed cancer-proofing mechanisms too. Its major cancer weapon is the interferon-mediated cell death mechanism. The gene for a certain form of interferon is duplicated compared to the one found in the common mouse or the naked mole rat [47].

Grey squirrels display abundant telomerase, yet when their cells are placed in culture dish, their spontaneous proliferation is surprisingly slow [47].

Moving further, the *Balaena mysticetus* **bowhead whale** is the longest-lived mammal known [91]. It is estimated to live around 200 years [91]. The animal shows signs of gradual senescence much later than humans. Given the enormous cell division it takes to reach adulthood when weighing 75-100 tonnes, it is amazing they rarely succumb to cancer, one of the frequent age-related diseases [91]. The bowhead whale is the only baleen whale to spend its whole life in Arctic freezing waters.

As I previously mentioned, around 90% of human tumors are able to divide relentlessly by reactivating or upregulating telomerase. The ALT pathway is detected in few cancers such as sarcomas. The main fear in activating telomerase in somatic human tissues is unleashing cancer.

Yet species with indeterminate growth expressing telomerase in their adult somatic tissues rarely suffer from cancer. This is the case of **decapod crustaceans** like lobsters, crabs, shrimp and crayfish [46].

Could this be a lack of information more than a low cancer rate? I doubt it. This is a commercially important group of animals for which huge amounts of data exist. Secondly, these animals are highly exposed to carcinogens being benthic species living at the bottom of the water. They maintain their stem cells in good stand by the end of their lives. Their only immune system is innate and it is able to encapsulate all sorts of foreign material, including cancer cells [46].

All these facts makes decapod crustaceans excellent models of cancer-resistant species. Other species in which cancer rates are very low are **sea urchins** and **long-lived seabirds** [46]. Apart from their regenerative abilities, **axolotls** are known for their cancer resistance as well [97].

Several preventative or curative cancer treatments can be developed from studying the genetic strategies employed by long-lived animals. Could larger weight hyaluronan be one day used in people? Could its turnover be slowed down by blocking the enzyme that breaks down this substance? Humans have already benefited from a genetic strategy invented by fungi: antibiotics saved countless lives. It is now time to learn other kinds of genetic strategies that may one day make cancer a thing of the past.

The End

Aging is a plastic phenomenon. Lifespan differences among species are orders of magnitude larger than any lifespan variation achieved in the lab. This is the reason for which I studied countless information resources in an attempt to gather highly specialized research into one easy-to-follow book. I intentionally wrote this book in plain English. Aging research is too important to hide it behind the closed doors of formal scientific jargon.

Gerontology as a science can progress by studying not only short-lived species like mice and worms, but gradually and especially negligibly senescent ones like sponges, naked mole rats, sea urchins, olms and many millennial trees. If aging is an increase in mortality rates and a decrease in fertility rates, then the existence of negligible senescence species indirectly shows that aging is an accident of nature.

Exceptionally long-lived species are characterized by:

- an environment low in predators, but high in climatic stresses
- slow growth even when exhibiting indeterminate growth
- late onset of maturation
- a primitive innate immune system
- partial or complete regeneration abilities as adults
- excellent protein control and gene maintenance over the whole lifespan
- cell membranes which are highly resistant to peroxidation

Climatic stresses like low temperature associated with oxygen and calorie restriction favored the evolution of neoteny – the preservation of juvenile traits along with gonad maturation. Instead of suffering reoxygenation-reperfusion injury like humans do during prolonged hypoxia and ischemia, neotenic species developed an extreme tolerance to anoxia and starvation.

Long-lived species often continue to express telomerase in their adult somatic tissues allowing them to regenerate at least part of their organs. Despite their adult expression of telomerase, such species do not have a higher rate of cancer. They probably evolved alternative mechanisms to keep cancer at bay while increasing the contact sensitivity of their cells. The naked mole rat is considered a cancer-proof species, despite abundant telomerase expression in its somatic stem cells.

Depressed metabolism during the earliest stages is able to prolong the total lifespan up to a point.

Slow growth is present in many long-lived species where growth hormones are inhibited from acting at their previous juvenile levels. The same kind of hormones wreak havoc in accelerated aging species like the Pacific salmon or the male marsupial mouse.

The magnitude of the project makes this book a work in progress. There are still countless species to be discovered. There are still aging experiments to be done and theories to be created. Aging is an accident of nature. And gerontology gerontology – the science of aging – was born to solve the puzzle of aging. Hopefully the book you are now holding gave you the promised insights regarding the aging gap between species.

Acknowledgments

This book would be lost of meaning if it didn't reach you, my dear reader. And it would have been impossible to create if green tea, public libraries and the Internet didn't exist.

Writing a book is like running a marathon. The other half of my soul, Daniel Ivan, played a key role in bringing this book into your hands. He provided me with plenty of support when I got overwhelmed with this project. When I got too focused for hours in a row, he baked vegetables for me and reminded me to eat. In exchange for a delicious boat of sushi, he provided insightful suggestions and appreciated criticism when the manuscript was still in its infancy. My dear friend Laura Dierksmeier answered all my strange English grammar questions when I wasn't certain how certain expressions would sound to a native ear. All the remaining errors are entirely mine.

Several people recommended me interesting papers to read: James Wm Clement, Marios Kyriazis, Mihaela Poteraşu and Paul Sandford McGlothin. Without such recommendations, this project would have been lacking in breadth and depth.

I am most indebted to my parents who decided to not only bring a new life on this blue planet, but to support it at every step possible too.

Yes, you made it. You reached the end of this book. Thank you for being with me during this adventure.

Anca Ioviţă

Acknowledgments

Bibliography

1. Abele, Doris, Jose Pablo Vazquez-Medina, and Tania Zenteno-Savin. "Aging in marine animals." In *Oxidative Stress in Aquatic Ecosystems*. Hoboken, NJ: Wiley-Blackwell, 2012.

2. André, S., F. Kerfourn, P. Affaticati, A. Guerci, and P. Ravassard. "Highly Restricted Diversity of TCR Delta Chains of the Amphibian Mexican Axolotl (Ambystomamexicanum) in Peripheral Tissues." *European Journal of Immunology* 37, no. 6 (2007): 1621-1633. doi:10.1002/eji.200636375.

3. Arendt, J. D., and D. N. Reznick. "Evolution of Juvenile Growth Rates in Female Guppies (Poecilia Reticulata): Predator Regime or Resource Level?" *Proc Biol Sci.* 272, no. 1560 (February 2005): 333-337. doi:10.1098/rspb.2004.2899.

4. Austad, S. N. "Retarded Senescence in an Insular Population of Virginia Opossums (Didelphis Virginiana)." *Journal of Zoology* 229, no. 4 (April 1993): 695-708. doi:10.1111/j.1469-7998.1993.tb02665.x.

5. Bidder, G. P. "Senescence." *British Medical Journal* 2 (1932): 583-585. doi:10.1136/bmj.2.3742.583.

6. Biga, P. R., and F. W. Goetz. "Zebrafish and Giant Danio As Models for Muscle Growth: Determinate Vs. Indeterminate Growth As Determined by Morphometric Analysis." *American Journal of Physiology-regulatory Integrative and Comparative Physiology* 291, no. 5 (November 2006): 1327-1337. doi:10.1152/ajpregu.00905.2005.

7. Biliński, T., T. Paszkiewicz, and R. Zadrag-Tecza. "Energy excess is the main cause of accelerated aging of mammals." *Oncotarget* 30, no. 6 (May 2015): 12909-19.

8. Blagosklonny, MV. "Aging and immortality: quasi-programmed senescence and its pharmacologic inhibition." *Cell Cycle* 5, no. 18 (September 2006): 2087-2102.

9. Blagosklonny, MV. "Once again on rapamycin-induced insulin resistance and longevity: despite of or owing to." *Aging* 4, no. 5 (May 2012): 350-8.

10. Bodnar, A. G. "Cellular and molecular mechanisms of negligible senescence: insight from the sea urchin." *Invertebr Reprod Dev* 59, no. 1 (January 2015): 23-27. doi:10.1080/07924259.2014.938195.

11. Boehm, AM, K. Khalturin, F. Anton-Erxleben, G. Hemmrich, UC Klostermeier, JA Lopez-Quintero, HH Oberg, et al. "FoxO is a critical regulator of stem cell maintenance in immortal Hydra." *Proc Natl Acad Sci U S A.* 109, no. 48 (November 2012): 19697-702. doi:10.1073/pnas.1209714109.

12. Brümmer, F., M. Pfannkuchen, A. Baltz, T. Hauser, and V. Thiel. "Light inside sponges." *Journal of Experimental Marine Biology and Ecology* 367, no. 2 (December 2008): 61-64. doi:10.1016/j.jembe.2008.06.036.

13. Buchmann, K. "Evolution of innate immunity: clues from invertebrates via fish to mammals." *Frontiers in immunology* 5 (September 2014): 459. doi:10.3389/fimmu.2014.

14. Buffenstein, R. "The Naked Mole-Rat: A New Long-Living Model

for Human Aging Research." *J Gerontol A Biol Sci Med Sci* 60, no. 11 (November 2005): 1369-77.

15. Buffenstein, R., and M. Pinto. "Endocrine Function in Naturally Long-living Small Mammals." *Molecular and Cellular Endocrinology* 299, no. 1 (February 2009): 101-111. doi:10.1016/j.mce.2008.04.021.

16. Buffenstein, R. "Negligible Senescence in the Longest Living Rodent, the Naked Mole-rat: Insights from a Successfully Aging Species." *J Comp Physiol B* 178, no. 4 (May 2008): 439-45. doi:10.1007/s00360-007-0237-5.

17. Buffenstein, R., R. Woodley, C. Thomadakis, TJ Daly, and DA Gray. "Cold-induced changes in thyroid function in a poikilothermic mammal, the naked mole-rat." *Am J Physiol Regul Integr Comp Physiol.* 280, no. 1 (January 2001): R149-55.

18. Bullard, K. M., M. T. Longaker, and H. P. Lorenz. "Fetal wound healing: current biology." *World Journal of Surgery* 27, no. 1 (January 2003): 54-61. doi:10.1007/s00268-002-6737-2.

19. Casselman, A. "Strange but true: the largest organism on earth is a fungus." *Scientific American*, October 4, 2007.

20. Caviedes-Vidal, E., W. H. Karasov, J. G. Chediack, V. Fasulo, A. P. Cruz-Neto, and L. Otani. "Paracellular Absorption: A Bat Breaks the Mammal Paradigm." *PLoS One* 3, no. 1 (January 2008): e1425. doi:10.1371/journal.pone.0001425.t002.

21. Clark, William R. *In Defense of Self: How the Immune System Really Works*. New York: Oxford University Press, 2008.

22. Congdon, J. D., R. D. Nagle, O. M. Kinney, and R. C. Sels. "Hypotheses of Aging in a Long-lived Vertebrate, Blanding's Turtle (Emydoidea Blandingii)." *Experimental Gerontology* 36, no. 4 (April 2001): 813-827. doi:10.1016/S0531-5565(00)00242-4.

23. Congdon, J. D., R. D. Nagle, O. M. Kinney, R. C. Sels, T. Quinter, and D. W. Tinkle. "Testing Hypotheses of Aging in Long-lived Painted Turtles (Chrysemys Picta)." *Experimental Gerontology* 38, no. 7 (July 2003): 765-72. doi:10.1016/S0531-5565(03)00106-2.

24. Dang, CV. "A metabolic perspective of Peto's paradox and cancer." *Philos Trans R Soc Lond B Biol Sci* 370, no. 1673 (July 2015). doi:10.1098/rstb.2014.0223.

25. Danilova, N., V. S. Hohman, F. Sacher, T. Ota, C. E. Willett, and L. A. Steiner. "T Cells and the Thymus in Developing Zebrafish." *Developmental and Comparative Immunology* 28, no. 7-8 (July 2004): 755-67. doi:10.1016/j.dci.2003.12.003.

26. De Grey, Aubrey, and Michael Rae. *Ending Aging: The Rejuvenation Breakthroughs That Could Reverse Human Aging in Our Lifetime.* New York: St. Martin's Press, 2007.

27. De Jesus, B., E. Vera, K. Schneeberger, AM Tejera, E. Ayuso, F. Bosch, and MA Blasco. "Telomerase gene therapy in adult and old mice delays aging and increases longevity without increasing cancer." *EMBO Mol Med* 4, no. 8 (August 2012): 691-704. doi:10.1002/emmm.201200245.

28. Delaney, R. G., S. Lahiri, and A. P. Fishman. "Aestivation of the African lungfish Protopterus aethiopicus: cardiovascular and respiratory functions." *J Exp Biol* 61, no. 1 (August 1974): 111-28.

29. Dmitriew, C. M. "The Evolution of Growth Trajectories: What Limits Growth Rate?" *Biol Rev Camb Philos Soc* 86, no. 1 (February 2011): 97-116. doi:10.1111/j.1469-185X.2010.00136.x.

30. Domínguez-Gerpe, L., and M. Rey-Méndez. "Evolution of the Thymus Size in Response to Physiological and Random Events Throughout Life." *Microscopy Research and Technique* 62, no. 6 (November 2003): 464-76. doi:10.1002/jemt.10408.

31. Duffy, J. E. "Species Boundaries, Specialization, and the Radiation of Sponge-dwelling Alpheid Shrimp." *Biological Journal of The Linnean Society* 58, no. 3 (1996): 307-324. doi:10.1111/j.1095-8312.1996.tb01437.x.

32. Ewers, F. W., and R. Schmid. "Longevity of Needle Fascicles of Pinus Longaeva (Bristlecone Pine) and Other North American Pines." *Oecologia* 59, no. 1 (January 1981): 23-27. doi:10.1007/BF00344660.

33. Ferguson, B. A., T. A. Dreisbach, C. G. Parks, G. M. Filip, and C. L. Schmitt. "Coarse-scale population structure of pathogenic Armillaria species in a mixed-conifer forest in the Blue Mountains of northeast Oregon." *Canadian Journal of Forest Research-revue Canadienne De Recherche Forestiere* 33, no. 4 (2003): 612-623. doi:10.1139/x03-065.

34. Finch, Caleb. *Longevity, Senescence, and the Genome*. Chicago: University of Chicago Press, 1990.

35. Forsyth, N. R., W. E. Wright, and J. W. Shay. "Telomerase and Differentiation in Multicellular Organisms: Turn It Off, Turn It On, and Turn It off Again." *Differentiation* 69, no. 4-5 (January 2002):

188-97. doi:10.1046/j.1432-0436.2002.690412.x.

36. Francis, N., T. Gregg, R. Owen, T. Ebert, and A. Bodnar. "Lack of Age-associated Telomere Shortening in Long- and Short-lived Species of Sea Urchins." *FEBS Letters* 580, no. 19 (August 2006): 4713-7. doi:10.1016/j.febslet.2006.07.049.

37. Freeman, HJ. ""Melanosis" in the small and large intestine." *World Journal of Gastroenterology* 14, no. 27 (July 2008): 4296-9. doi:10.3748/wjg.14.4296.

38. Funayama, N. "The stem cell system in demosponges: suggested involvement of two types of cells: archeocytes (active stem cells) and choanocytes (food-entrapping flagellated cells)." *Development genes and evolution* 223, no. 1-2 (March 2013): 23-38. doi:10.1007/s00427-012-0417-5.

39. Gaino, E., G. Bavestrello, and G. Magnino. "Self/non-self recognition in sponges." *Italian Journal of Zoology* 66, no. 4 (1999): 299-315.

40. Gardner, M. P., D. Gems, and M. E. Viney. "Extraordinary Plasticity in Aging in Strongyloides Ratti Implies a Gene-regulatory Mechanism of Lifespan Evolution." *Aging Cell* 5, no. 4 (August 2006): 315-23. doi:10.1111/j.1474-9726.2006.00226.x.

41. Gatti, S. "The role of sponges in high-Antarctic carbon and silicon cycling: a modelling approach." *Alfred-Wegener-Inst. f. Polar-u.Meeresforschung*,2002.
http://epic.awi.de/26613/1/BerPolarforsch2002434.pdf.

42. Gavrilov, L. A., and N. S. Gavrilova. "The Reliability Theory of

Aging and Longevity." *Journal of Theoretical Biology* 213, no. 4 (December 2001): 527-45. doi:10.1006/jtbi.2001.2430.

43. George JC et al. "Energetic adaptations of the bowhead whales." *Abstracts of 14th marine mammal biennial conference, Vancouver, Canada*, 2001.

44. Gilbert, Scott F. "Metamorphosis, regeneration, and aging." In *Developmental Biology*, 6th ed. Sunderland, Mass.: Sinauer Associates, 2000.

45. Godwin, J. W., A. R. Pinto, and N. A. Rosenthal. "Macrophages are required for adult salamander limb regeneration." *Proceedings of the National Academy of Sciences* 110, no. 23 (2013): 9415-9420. doi:10.1073/pnas.1300290110.

46. Gomes, N. M., J. W. Shay, and W. E. Wright. "Telomere Biology in Metazoa." *FEBS Lett* 584, no. 17 (September 2010): 3741-3751. doi:10.1016/j.febslet.2010.07.031.

47. Gorbunova, V., A. Seluanov, Z. Zhang, VN Gladyshev, and J. Vijg. "Comparative genetics of longevity and cancer: insights from long-lived rodents." *Nat Rev Genet* 15, no. 8 (August 2014): 531-40. doi:10.1038/nrg3728.

48. Gruber, H., R. Schaible, ID Ridgway, TT Chow, C. Held, and EE Philipp. "Telomere-independent ageing in the longest-lived non-colonial animal, Arctica islandica." *Exp Gerontol* 51 (March 2014): 38-45. doi:10.1016/j.exger.2013.12.014.

49. Guidetti, R., T. Altiero, and L. Rebecchi. "On dormancy strategies in tardigrades." *Journal of Insect Physiology* 57, no. 5 (May 2011):

567-576. doi:10.1016/j.jinsphys.2011.03.003.

50. Haddad, L. S., L. Kelbert, and A. J. Hulbert. "Extended Longevity of Queen Honey Bees Compared to Workers is Associated with Peroxidation-resistant Membranes." *Exp Gerontology* 42, no. 7 (July 2007): 601-9. doi:10.1016/j.exger.2007.02.008.

51. Hariharan, IK, DB Wake, and MH Wake. "Indeterminate Growth: Could It Represent the Ancestral Condition?" *Cold Spring Harb Perspect Biol*, July 2015. doi:10.1101/cshperspect.a019174. [Epub ahead of print].

52. Heino, M., and V. Kaitala. "Evolution of Resource Allocation Between Growth and Reproduction in Animals with Indeterminate Growth." *Journal of Evolutionary Biology* 12, no. 3 (1999): 423-429. doi:10.1046/j.1420-9101.1999.00044.x.

53. Hervant, F., J. Mathieu, and J. Durand. "Behavioural, physiological and metabolic responses to long-term starvation and refeeding in a blind cave-dwelling (Proteus anguinus) and a surface-dwelling (Euproctus asper) salamander." *J Exp Biol* 204, no. 2 (January 2001): 269-81.

54. "Hexactinellid Sponge (Scolymastra Joubini) Longevity, Ageing, and Life History." Human Ageing Genomic Resources. Accessed September 7, 2015. http://genomics.senescence.info/species/entry.php?species=Scolymastra_joubini.

55. Hulbert, A. J., R. Pamplona, R. Buffenstein, and W. A. Buttemer. "Life and Death: Metabolic Rate, Membrane Composition, and Life Span of Animals." *Physiological Reviews* 87, no. 4 (October 2007): 1175-213. doi:10.1152/physrev.00047.2006.

56. Jackson, D. C. "Hibernating Without Oxygen: Physiological Adaptations of the Painted Turtle." *J Physiol* 543, no. 3 (September 2002): 731-737. doi:10.1113/jphysiol.2002.024729.

57. Jemielity, S., M. Chapuisat, JD Parker, and L. Keller. "Long live the queen: studying aging in social insects." *Age* 27, no. 3 (September 2005): 241-8. doi:10.1007/s11357-005-2916-z.

58. Jerne, Niels Kaj, and Ivan Lefkovits. *A Portrait of the Immune System Scientific Publications of N.K. Jerne*. Singapore: World Scientific, 1996.

59. Jochum, KP, X. Wang, TW Vennemann, B. Sinha, and WE Müller. "Siliceous deep-sea sponge Monorhaphis chuni: A potential paleoclimate archive in ancient animals." *Chemical Geology* 300 (March 2012): 143-151. doi:10.1016/j.chemgeo.2012.01.009.

60. Jönsson, K. I., E. Rabbow, R. O. Schill, M. Harms-Ringdahl, and P. Rettberg. "Tardigrades survive exposure to space in low Earth orbit." *Current Biology* 18, no. 17 (September 2008): R729-R731. doi:10.1016/j.cub.2008.06.048.

61. Kara, TC. "Ageing in Amphibians." *Gerontology* 40, no. 2-4 (1994): 161-73. doi:10.1159/000213585.

62. Kawamoto, K. "Endocrine Control of the Reproductive Activity in Hibernating Bats." *Zoolog Sci* 20, no. 9 (September 2003): 1057-69. doi:10.2108/zsj.20.1057.

63. Keil, G., E. Cummings, and JP De Magalhães. "Being cool: how body temperature influences ageing and longevity." *Biogerontology*

16, no. 4 (August 2015): 383-97. doi:10.1007/s10522-015-9571-2.

64. Kishi, S. "Functional Aging and Gradual Senescence in Zebrafish." *Annals of The New York Academy of Sciences* 1019 (June 2004): 521-6. doi:10.1196/annals.1297.097.

65. Kozlowski, J., M. Czarnoleski, and M. Danko. "Can Optimal Resource Allocation Models Explain Why Ectotherms Grow Larger in Cold?" *Integrative and Comparative Biology* 44, no. 6 (December 2004): 480-93. doi:10.1093/icb/44.6.480.

66. Kuro-O, M., Y. Matsumura, H. Aizawa, H. Kawaguchi, T. Suga, T. Utsugi, Y. Ohyama, et al. "Mutation of the Mouse Klotho Gene Leads to a Syndrome Resembling Ageing." *Nature* 390, no. 6655 (November 1997): 45-51. doi:10.1038/36285.

67. Kurosu, H., M. Yamamoto, J. D. Clark, J. V. Pastor, A. Nandi, P. Gurnani, O. P. McGuinness, et al. "Suppression of Aging in Mice by the Hormone Klotho." *Science* 309, no. 5742 (September 2005): 1829-1833. doi:10.1126/science.1112766.

68. Lane, Nick. *Power, Sex, Suicide: Mitochondria and the Meaning of Life*. Oxford; New York: Oxford University Press, 2005.

69. Larson, D. W. "The Paradox of Great Longevity in a Short-lived Tree Species." *Experimental Gerontology* 36, no. 4-6 (April 2001): 651-73. doi:10.1016/S0531-5565(00)00233-3.

70. Lee, J., S. Giordano, and J. Zhang. "Autophagy, mitochondria and oxidative stress: cross-talk and redox signalling." *Biochemical Journal* 441, no. 2 (January 2012): 523-540. doi:10.1042/BJ20111451.

71. Leri, A., S. Franco, A. Zacheo, L. Barlucchi, S. Chimenti, F. Limana, B. Nadal-Ginard, et al. "Ablation of telomerase and telomere loss leads to cardiac dilatation and heart failure associated with p53 upregulation." *EMBO Journal* 22, no. 1 (January 2003): 131-9. doi:10.1093/emboj/cdg013.

72. Lewington, Anna, and Edward Parker. *Ancient Trees: Trees That Live for a Thousand Years*. London: Batsford, 2012.

73. Lika, K., and S. A. Kooijman. "Life History Implications of Allocation to Growth Versus Reproduction in Dynamic Energy Budgets." *Bull Math Biol* 65, no. 5 (September 2003): 809-34. doi:10.1016/S0092-8240(03)00039-9.

74. Lund, T. C., T. J. Glass, J. Tolar, and B. R. Blazar. "Expression of telomerase and telomere length are unaffected by either age or limb regeneration in Danio rerio." *PLOS One* 4, no. 11 (November 2009): e7688. doi:10.1371/journal.pone.0007688.

75. Maciak, S., and P. Michalak. "Cell size and cancer: a new solution to Peto's paradox?" *Evolutionary applications* 8, no. 1 (January 2015): 2-8. doi:10.1111/eva.12228.

76. Magalhaes, J. P., J. Costa, and G. M. Church. "An Analysis of the Relationship Between Metabolism, Developmental Schedules, and Longevity Using Phylogenetic Independent Contrasts." *J Gerontol A Biol Sci Med Sci* 62, no. 2 (2007): 149-160.

77. Miller, J. K. "Escaping Senescence: Demographic Data from the Three-toed Box Turtle (Terrapene Carolina Triunguis)." *Experimental Gerontology* 36, no. 4-6 (April 2001): 829-32. doi:10.1016/S0531-5565(00)00243-6.

78. Mitchell Harman, S. "Endocrine Changes with Aging." Evidence-Based Clinical Decision Support at the Point of Care | UpToDate. Last modifiedNovember7,2014. http://www.uptodate.com/contents/endocrine-changes-with-aging.

79. Moon, Beth. *Ancient Trees: Portraits of Time*. New York: Abbeville Press Publishers, 2014.

80. Morselli, E., M. C. Maiuri, M. Markaki, E. Megalou, A. Pasparaki, K. Palikaras, A. Criollo, et al. "Caloric restriction and resveratrol promote longevity through the Sirtuin-1-dependent induction of autophagy." *Cell Death and Disease* 1, no. 1 (January 2010): e10. doi:10.1038/cddis.2009.8.

81. Munné-Bosch, S. "Perennial roots towards immortality." *Plant Physiology Preview* 166, no. 2 (February 2014): 720-725. doi:10.1104/pp.114.236000.

82. Murray, N. E. "2001 Fred Griffith review lecture. Immigration control of DNA in bacteria: self versus non-self." *Microbiology* 148, no. 1 (January 2002): 3-20. doi:10.1099/00221287-148-1-3.

83. Nagy, J. D., E. M. Victor, and J. H. Cropper. "Why don't all whales have cancer? A novel hypothesis resolving Peto's paradox." *Integrative and Comparative Biology* 47, no. 2 (August 2007): 317-28. doi:10.1093/icb/icm062.

84. Osiewacz, H. D. "Genes, mitochondria and aging in filamentous fungi." *Ageing Research Reviews* 1, no. 3 (June 2002): 425-42. doi:10.1016/S1568-1637(02)00010-7.

85. Ottenweller, J. E., W. N. Tapp, D. L. Pitman, and B. H. Natelson. "Interactions among the effects of aging, chronic disease, and stress on adrenocortical function in Syrian hamsters." *Endocrinology* 126, no. 1 (January 1990): 102-9. doi:10.1210/endo-126-1-102.

86. Page Jr., RE, and CY Peng. "Aging and development in social insects with emphasis on the honey bee, Apis mellifera L." *Experimental Gerontology* 36, no. 4-6 (April 2001): 695-711. doi:10.1016/S0531-5565(00)00236-9.

87. Patnaik, B. K., N. Mahapatro, and B. S. Jena. "Ageing in Fishes." *Gerontology* 40, no. 2-4 (1994): 113-32. doi:10.1159/000213582.

88. Petralia, RS, MP Mattson, and PJ Yao. "Aging and longevity in the simplest animals and the quest for immortality." *Ageing research reviews* 16 (July 2014): 66-82. doi:10.1016/j.arr.2014.05.003.

89. Piraino, S., F. Boero, B. Aeschbach, and V. Schmid. "Reversing the life cycle: medusae transforming into polyps and cell transdifferentiation in Turritopsis nutricula (Cnidaria, Hydrozoa)." *Biological Bulletin*, 1996, 302-312.

90. Pride, H., Z. Yu, B. Sunchu, J. Mochnick, A. Coles, Y. Zhang, R. Buffenstein, PJ Hornsby, SN Austad, and VI Pérez. "Long-lived species have improved proteostasis compared to phylogenetically-related shorter-lived species." *Biochem Biophys Res Commun* 457, no. 4 (February 2015): 669-75. doi:10.1016/j.bbrc.2015.01.046.

91. Prokopov, A. F. "Theoretical paper: exploring overlooked natural mitochondria-rejuvenative intervention: the puzzle of bowhead whales and naked mole rats." *Rejuvenation Research* 10, no. 4 (December 2007): 543-60. doi:10.1089/rej.2007.0546.

92. Reznick, D., M. Bryant, and D. Holmes. "The evolution of senescence and post-reproductive lifespan in guppies (Poecilia reticulata)." *PLOS Biol* 4, no. 1 (January 2006): e7. doi:10.1371/journal.pbio.0040007.

93. Reznick, D., C. Ghalambor, and L. Nunney. "The evolution of senescence in fish." *Mech Ageing Dev* 123, no. 7 (April 2002): 773-89. doi:10.1016/S0047-6374(01)00423-7.

94. Rinkevich, B., and Y. Loya. "Senescence and dying signals in a reef building coral." *Experientia* 42, no. 3 (1986): 320-322. doi:10.1007/BF01942521.

95. Roark, E. B., T. P. Guilderson, R. B. Dunbar, S. J. Fallon, and D. A. Mucciarone. "Extreme longevity in proteinaceous deep-sea corals." *Proceedings of The National Academy of Sciences* 106, no. 13 (March 2009): 5204-8. doi:10.1073/pnas.0810875106.

96. "Rocky Mountain Tree-Ring Research, OLDLIST." Rocky Mountain Tree-Ring Research Home Page. Last modified January 2013. http://www.rmtrr.org/oldlist.htm.

97. Roy, S., and S. Gatien. "Regeneration in Axolotls: a Model to Aim For!" *Exp Gerontol* 43, no. 11 (November 2008): 968-973. doi:10.1016/j.exger.2008.09.003.

98. Ruppert, Edward E., Richard S. Fox, and Robert D. Barnes. *Invertebrate Zoology: A Functional Evolutionary Approach*, 7th ed. Belmont, California: Thomson-Brooks/Cole, 2004.

99. Schmich, J., Y. Kraus, D. D. Vito, D. Graziussi, F. Boero, and S.

Piraino. "Induction of reverse development in two marine Hydrozoans." *International Journal of Developmental Biology* 51, no. 1 (2007): 45-56. doi:10.1387/ijdb.062152js.

100. Schmidt, MH. "The energy allocation function of sleep: a unifying theory of sleep, torpor, and continuous wakefulness." *Neurosci Biobehav Rev* 47 (November 2014): 122-53. doi:10.1016/j.neubiorev.2014.08.001.

101. Sinensky, M. "Homeoviscous adaptation--a homeostatic process that regulates the viscosity of membrane lipids in Escherichia coli." *Proc Natl Acad Sci* 71, no. 2 (February 1974): 522-525. doi:10.1073/pnas.71.2.522.

102. Solana, J. "Closing the circle of germline and stem cells: the primordial stem cell hypothesis." *EvoDevo* 4, no. 1 (January 2013): 2. doi:10.1186/2041-9139-4-2.

103. "Species with Negligible Senescence." Human Ageing Genomic Resources.AccessedSeptember 7, 2015. http://genomics.senescence.info/species/nonaging.php.

104. Storey, K. B., J. M. Storey, S. P. Brooks, T. A. Churchill, and R. J. Brooks. "Hatchling turtles survive freezing during winter hibernation." *Proceedings of The National Academy of Sciences of the USA* 85, no. 21 (November 1988): 8350-4. doi:10.1073/pnas.85.21.8350.

105. Strahl, Julia. "Life strategies in the long-lived bivalve Arctica islandica on a latitudinal climate gradient – Environmental constraints and evolutionary adaptations." Master's thesis, Universität Bremen, 2011.

106. Sussman, Rachel, Hans Ulrich Obrist, Carl Zimmer, Christina Louise Costello, and Michael Paukner. *The Oldest Living Things in the World*. Chicago; London: University of Chicago Press, 2014.

107. Sutherland, JS, GL Goldberg, MV Hammett, AP Uldrich, SP Berzins, TS Heng, BR Blazar, et al. "Activation of thymic regeneration in mice and humans following androgen blockade." *Journal of immunology* 175, no. 4 (August 2005): 2741-53.

108. Tan, TC, R. Rahman, F. Jaber-Hijazi, DA Felix, C. Chen, EJ Louis, and A. Aboobaker. "Telomere maintenance and telomerase activity are differentially regulated in asexual and sexual worms." *Proc Natl Acad Sci U S A* 109, no. 11 (March 2012): 4209-14. doi:10.1073/pnas.1118885109.

109. Trumble, SJ, EM Robinson, M. Berman-Kowalewskic, CW Potterd, and S. Usenko. "Blue whale earplug reveals lifetime contaminant exposure and hormone profiles." *Proceedings of the National Academy of Sciences* 110, no. 42 (August 2013): 16922–16926.

110. Turbill, C., C. Bieber, and T. Ruf. "Hibernation is associated with increased survival and the evolution of slow life histories among mammals." *Proceedings of the Royal Society of London B: Biological Sciences* 278, no. 1723 (2011): 3355-3363. doi:10.1098/rspb.2011.0190.

111. Uribarri, J., S. Woodruff, S. Goodman, W. Cai, X. Chen, R. Pyzik, A. Yong, G. E. Striker, and H. Vlassara. "Advanced glycation end products in foods and a practical guide to their reduction in the diet." *Journal of The American Dietetic Association* 110, no. 6 (June 2010): 911-16.e12.. doi:10.1016/j.jada.2010.03.018.

112. Vaupel, J. W., A. Baudisch, M. Dölling, D. A. Roach, and J. Gampea. "The case for negative senescence." *Theoretical Population Biology* 65, no. 4 (June 2004): 339-351. doi:10.1016/j.tpb.2003.12.003.

113. Voituron, Y., M. De Fraipont, J. Issartel, O. Guillaume, and J. Clobert. "Extreme lifespan of the human fish (Proteus anguinus): a challenge for ageing mechanisms." *Biology Letters* 7, no. 1 (February 2011): 105-7. doi:10.1098/rsbl.2010.0539.

114. Wagner, D. E., I. E. Wang, and P. W. Reddien. "Clonogenic neoblasts are pluripotent adult stem cells that underlie planarian regeneration." *Science* 332, no. 6031 (May 2011): 811-6. doi:10.1126/science.1203983.

115. Wakahara, M. "Heterochrony and neotenic salamanders: possible clues for understanding the animal development and evolution." *Zoological Science* 13, no. 6 (December 1996): 765-76. doi:10.2108/zsj.13.765.

116. Wang-Michelitsch, J., and TM Michelitsch. "Development of age spots as a result of accumulation of aged cells in aged skin." May 2015. arXiv preprint arXiv:1505.07012.

117. Yang, T., and R. Buffenstein. "Effect of aging on glycated hemoglobin and blood glucose concentration in naked mole-rats." *FASEB JOURNAL* 18, no. 5 (March 2004): A1301-A1302.

Bibliography